FELIX BEREZIN

LIFE AND DEATH OF THE MASTERMIND
OF SUPERMATHEMATICS

FELIX BEREZIN

LIFE AND DEATH OF THE MASTERMIND OF SUPERMATHEMATICS

Editor

M. SHIFMAN
University of Minnesota, USA

World Scientific

NEW JERSEY • LONDON • SINGAPORE • BEIJING • SHANGHAI • HONG KONG • TAIPEI • CHENNAI

Published by

World Scientific Publishing Co. Pte. Ltd.

5 Toh Tuck Link, Singapore 596224

USA office: 27 Warren Street, Suite 401-402, Hackensack, NJ 07601

UK office: 57 Shelton Street, Covent Garden, London WC2H 9HE

British Library Cataloguing-in-Publication Data
A catalogue record for this book is available from the British Library.

The Editor and Publisher would like to thank the American Mathematical Society for their kind permission to reprint the articles of V. Maslov, R. Minlos, M. Shubin, N. Vvedenskaya and the first article of A. Vershik from the *American Mathematical Society Translations*, Series 2, *Advances in the Mathematical Sciences*, Volume 175, 1996.

Paintings Alexandra Rozenman
Black-and-white sketches Yuri Korjevsky
Graphic design Leigh Simmons
Photographs Elena Karpel's collection

FELIX BEREZIN: LIFE AND DEATH OF THE MASTERMIND OF SUPERMATHEMATICS

ISBN-13 978-981-270-532-7
ISBN-10 981-270-532-5
ISBN-13 978-981-270-533-4 (pbk)
ISBN-10 981-270-533-3 (pbk)

Printed in Singapore by World Scientific Printers (S) Pte Ltd

FOREWORD

M. SHIFMAN
W.I. Fine Theoretical Physics Institute, University of Minnesota, Minneapolis, MN 55455, USA
shifman@umn.edu

The story of this Memorial Volume is as follows. In the fall of 2005 Arkady Vainshtein mentioned in passing that he had received Elena Karpel's essay on Felix Berezin from his friend Dmitri Gitman. Of course, every student and every practitioner of modern field theory knows the Berezin integral over the Grassmann variables, which constitutes the basis of the current approach to theories with fermions and quantization of gauge theories (introduction of ghosts). Without using the Berezin integral, string theory and supersymmetry studies, which are at the focus of modern high-energy physics, would be extremely hard, if not impossible. That was about the only thing I knew about Berezin, in addition to the fact that he had died in an accident in 1980 at the age of 49. I met him in person only once or twice in the late 1970s at seminars in Moscow. My impression was that the Berezin integral had always existed, "from the days of antiquity." I am sure, this is also the impression of young people who came to mathematical and theoretical physics in the 1990s and later.

Given Arkady's remark that the essay was very moving, and my interest in the history of high-energy physics, my immediate reaction was to get hold of this memoir as soon as possible and read it myself. I wrote to Gitman, and he kindly forwarded to me the original Russian text.

I read it right away. Indeed, it was both moving and masterfully written. To say that it was impressive is an understatement. I was captivated by the depth and literary maturity of Karpel's story, a story of the life and death of a remarkable man, an outstanding mathematician whose ideas were ahead of his time and in many instances were not immediately appreciated by his contemporaries. It was only later that they shaped entire areas of mathematical physics and proved to be instrumental in the development of modern high-energy theory and string theory.

What struck me most was an obvious similarity between Berezin's life and fate and those of other Soviet scholars on whose biographies I had worked in the past. This volume is my fourth book of this type. In 2000 I published two books honoring Yuri Golfand,[a] the theoretical physicist who discovered supersymmetry in four-dimensional field theories. In 2005 I edited a book dedicated to the memory of Bella Subbotovskaya,[b] a pioneer in the mathematics of complex systems and the founder of the Jewish People's University in Moscow in the late 1970s and early '80s. These three people were contemporaries, but everything else was different — age, gender, family status, character traits, general interests, areas in which they worked... And yet, they shared a common destiny: intellectual potential wasted in vain, suppressed talents, unacknowledged skills, unfulfilled expectations, everyday humiliations, neglect and outright persecution by the authorities... Their lives were broken, just as the lives of so many other people whose only "fault" was to be born in the wrong place at the wrong time. Officially, they were supposed to enjoy the "advantages of real socialism," Soviet style. And instead?.. How many discoveries could have been made? How many good deeds could have been done? We will never know.

My involvement in Berezin's story became a personal issue, even though I never knew this man personally. Soon my investigations led me to a group of mathematicians[c] who in 1996 edited the *F.A. Berezin Memorial Volume* issued by the American Mathematical Society.[d] In addition to mathematical treatises this Volume

---

[a]*The Many Faces of the Superworld*, Ed. M. Shifman, (World Scientific, Singapore, 2000); *The Supersymmetric World*, Ed. G. Kane and M. Shifman, (World Scientific, Singapore, 2000).

[b]*You Failed Your Math Test, Comrade Einstein*, Ed. M. Shifman, (World Scientific, Singapore, 2005).

[c]R.L. Dobrushin, R.A. Minlos, M.A. Shubin, A.M. Vershik.

[d]*American Mathematical Society Translations*, Series 2, *Advances in the Mathematical Sciences*, Volumes 175 and 177 (AMS, Providence, Rhode Island, 1996).

contained recollections by some of Berezin's friends and colleagues. The reader will find these articles reprinted below, in Part II of the present book. I wrote to Misha Shubin, Anatoly Vershik and Nikita Vvedenskaya, who replied to my letters immediately promising their enthusiastic support. Professor Vershik went as far as to prepare a new essay on Felix Berezin, written expressly for this book, which you will also see in Part II. He also informed me of a recent Russian publication honoring Felix Berezin.[e]

In November of 2005 I got in touch with Elena Karpel, and in the summer of 2006 I met and interviewed her in La Rochelle, the French city where she now resides. In 1990 she and her daughter Natalie Berezin left the Soviet Union in a bid to start a new life in France. She told me of the fifteen years that had elapsed since then, a dramatic story of a long struggle for survival in an unfamiliar and often hostile environment. Being quite remarkable by itself, this story does not belong to this book, however. It may or may not be published elsewhere in the future.

Karpel's memoir, which constitutes the core of this book, is very dear to Elena Grigorievna. She tried first to publish the original Russian version in Russia. For various reasons this attempt failed, as did her subsequent attempt to publish a French translation in France. Thus, the English translation which opens Part II presents the first publication.

Elena Grigorievna connected me with Victor Palamodov who now resides in Israel and, somewhat later, with Dimitri Leites in Germany. They took this project close to their hearts. Not only did they contribute original articles, they also offered plenty of advice and comments as to possible improvements. I managed to implement some of them. From day one I was in permanent contact with Dmitri Gitman. His story of a few encounters with Felix Berezin presents a clear picture of the atmosphere we lived in. It nicely matches and, in a sense, supplements Karpel's memoir.

This book consists of three parts. Part I contains two surveys

[e] *Zapiski Nauchnyh Seminarov POMI*, Vol. 331, *Representation Theory, Dynamical Systems, Combinatorial and Algorithmic Methods*, Ed. A. Vershik, (St. Petersburg, 2006), pp. 1–235.

written by Andrei Losev and Robert Minlos summarizing the scientific contribution of Felix Berezin. They show Berezin's discoveries and innovative ideas in mathematical physics in a proper perspective, that of today's researcher.

Part II presents recollections by Elena Karpel and Berezin's colleagues, friends and students. Finally, in Part III of this Volume I publish the English translation of two documents from Berezin's archive: letters to the Rector of the Moscow State University and to the Governing Board of the Moscow Mathematical Society. Needless to say, currently they are only of historic interest. They acquaint the reader with the work environment to which Felix Berezin was confined. For the Western reader it will be hard to understand what he was fighting for. Ironically, the suggestions that Berezin made in his letters are standard procedure at Western universities.

In the process of conducting this work, I shared my excitement with many friends. Some of them had the opportunity to read Karpel's memoir either in its original form or at various stages of translation. One of the first readers was a well-known artist, Alexandra Rozenman, whose paintings are in high demand among the collectors and galleries. Captivated by the sad melody of this story she created the painting *One-Way Ticket* expressing this melody in a visual form. The reader will find a small reproduction in the beginning of Part II.

General acknowledgments

I am grateful to Ashleigh Geib who was responsible for typesetting a part of this volume in LaTeX. My deep gratitude goes to Anne Barthel for expert advice on English grammar. Her effort at improving the English translation of Karpel's memoir is impossible to overestimate. I want to thank Iman Benmachiche for her assistance with French, and Anya Dashevsky, Joel Giedt and Stephen Gasiorowicz for proofreading some parts of the manuscript. I am thankful to Roxanne Weymouth who handled all the financial aspects of this project.

My friend and colleague Dimitry Leites kindly agreed to proofread Losev's review and Part III of this book and made valuable remarks.

Special thanks go to Yuri Korjevsky, Alexandra Rozenman and

Leigh Simmons who worked hard to make this book visually appealing. They did a great job, creating several successive versions of the cover design and illustrations.

I am grateful to Roman Kovalev and Pavel Moshin who carried the main burden of translation. Regarding the translation of numerous poetic passages scattered in Karpel's memoir I consulted Professor Anatoly Liberman, to whom I would like to say thank you. I am grateful to Stan Lipovetsky for connecting me with Yuri Korjevsky in Moscow and assisting in subsequent communications with him. As usual, my World Scientific contact, Lakshmi Narayanan — "my Editor" — was instrumental in the speedy completion of this project.

This publication, which I hope will have a lasting impact in the community of mathematical physicists, was made possible by the enthusiastic participation of friends, colleagues and students of Felix Berezin from whom I received much advice and encouragement. Above all, I want to say *thank you* to the contributors who are listed below.

Additional acknowledgments

Paintings on the cover and pages 1 and 53 are by Alexandra Rozenman. The painting on the cover (*Where is Everybody?*) is from the collection of Caitlin Skinner and Robert Marlott. The painting on page 1 (*Again*) is courtesy of the artist. The painting on page 53 (*One-Way Ticket*) is from the M.&M. Shifman collection. Black-and-white sketches are by Yuri Korjevsky (M.&M. Shifman collection). All photographs are from the collection of Elena Karpel.

I thank the American Mathematical Society for their kind permission to reprint the articles of V. Maslov, R. Minlos, M. Shubin, N. Vvedenskaya and the first article of A. Vershik from the *American Mathematical Society Translations*, Series 2, *Advances in the Mathematical Sciences*, Volume 175, 1996.

January 16, 2007,
Minneapolis

List of contributors

Natalie Berezin, Dmitri Gitman, Elena Karpel, Dimitry Leites, Andrei Losev, Victor Maslov, Robert Minlos, Victor Palamodov, Misha Shubin, Anatoly Vershik, Nikita Vvedenskaya

Disclaimer: The opinions expressed by the writers herein are not necessarily those of the Editor.

CONTENTS

PART ONE

ANDREI LOSEV

ROBERT MINLOS

FROM BEREZIN INTEGRAL TO BATALIN–VILKOVISKY FORMALISM: A MATHEMATICAL PHYSICIST'S POINT OF VIEW

ANDREI LOSEV

Institute of Theoretical and Experimental Physics,
B. Cheremushkinskaya 25, 117259, Moscow, Russia
losev@itep.ru

Within the path integral approach quantum field theory can be formulated as an integral over the superspace of fields. The Berezin integral is the crucial element in this construction. Replacing an infinite-dimensional supermanifold of fields by an ordinary supermanifold we get a nice toy model of quantum field theory. In this article I motivate the origin of the Berezin integral and trace some of its applications, up to the Batalin–Vilkovisky formalism. The latter was developed as a method for determining the ghost structure in such theories as gravity and supergravity, whose Hamiltonian formalism has constraints not related to a Lie algebra action.

1. Instead of Introduction

1.1. *Mathematical physics as a style of theoretical physics*

You may like it or hate it, but during the last 40 years the shape of what was previously called theoretical physics has changed considerably. It has diversified in styles.

In the good old days, theoretizing was like sailing between islands of experimental evidence. And, if the trip was not in the vicinity of the shoreline (that was strongly recommended for safety reasons) sailors were continuously looking forward, hoping to see land — the sooner the better. Intellectual gamblers trying to cross the sea were mostly disappointed, lost and forgotten; lucky survivors were proclaimed geniuses, but — do not even think of doing it at home — no one would recommend this path for his friends or relatives.

Nowadays, some theoretical physicists (let us call them sailors) found a way to survive and navigate in the open sea of pure theoretical constructions. Instead of the horizon, they look at the stars,[a] which tell them exactly where they are. Sailors are aware of the fact that the stars will never tell them where the new land is, but they may tell them their position on the globe. In this way sailors — all together — are making a map that will at the end facilitate navigation in the sea and will help to discover a new land.

Theoreticians become *sailors* simply because they just like it. Young people seduced by captains forming crews to go to a Nuevo El Dorado of Unified Quantum Field Theory or Quantum Gravity soon realize that they will spend all their life at sea. Those who do not like sailing desert the voyage, but for true potential sailors the sea becomes their passion. They will probably tell the alluring and frightening truth to their students — and the proper people will join their ranks.

These sailors are known as mathematical physicists; they form one of the styles of modern theoretical physics. The author of this text considers himself to be one of them.

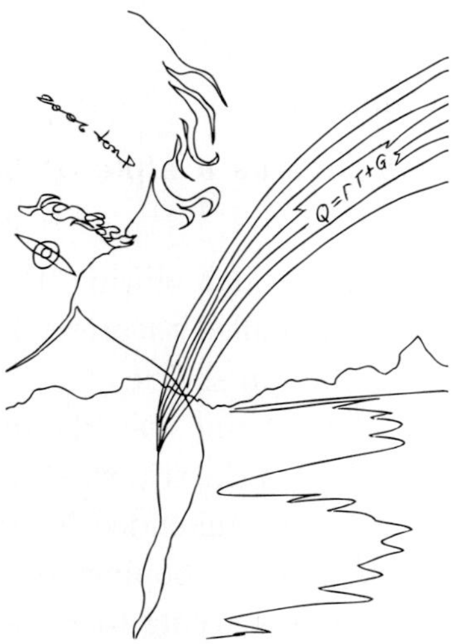

[a]Here by "stars" I mean internal logic organizing the mathematical world, rather than some outstanding members of the community.

Below I will explain my point of view on the *Berezin integral*, a truly remarkable navigating device. The reader may consider this an example of how the mathematical physicist approaches the world, or just share with the author admiration for the unexpected effectiveness of the Berezin integral in mathematical physics. Felix Berezin presented his "integral calculus" in anticommuting variables in his first book *Method of Second Quantization*, published in Russian in 1965 and translated into English in 1966 (see Appendix), which became a cherished handbook for generations of mathematical physicists.

I think that mathematical physics should study all possible quantum field theories (QFTs), by all available tools and methods. At first glance this is an incredibly hard problem, since one can hardly describe all possible QFTs, let alone solve them. However, they are partially ordered by their complexity.

Compare it to geometry which studies all possible spaces. The set of all spaces is intractable; however, the simplest spaces, such as a point, a line or a circle, are easy to imagine and understand. In real dimension 2 we can classify all smooth oriented spaces by genus. In higher dimensions things become increasingly more complicated.

I think that the main task of a mathematical physicist is to study the universal phenomena of QFT as functions on the space of all possible QFTs. In the simplest QFTs most phenomena are tautologies, like $0 = 0$. As complexity increases, interesting phenomena start showing up, at first in their simplest form. At this stage they are tame and their exhaustive quantitative description may be obtained. Increasing complexity makes typical phenomena described by the given QFT yet more complicated, and only an overall qualitative picture survives.

Thus, I assume that understanding the phenomena in QFT is equivalent to finding the simplest (threshold) QFT where nontrivial phenomena "start to fly."

1.2. *Outline of the paper*

One of the approaches to general QFT is the functional integral. It has to be taken over the superspace of fields. It means that some of these fields are fermionic (odd in mathematical terms). These fields include not only spinors corresponding to matter fields, such as electrons, but also ghost fields of gauge symmetries. There are "zero-dimensional" QFTs where the infinite-dimensional space of fields is replaced just by an ordinary finite-dimensional space, and the infinite-dimensional superspace of fields by a finite-dimensional superspace. The functional integral is then replaced by an ordinary integral. However, a part of quantum nature of the "theory" survives in a dependence on $\hbar$ that can be implemented even in this oversimplified model.

In what follows we will discuss the Berezin superspace integral, compare it to other integrals known in mathematics and re-express them in terms of the Berezin integral. We will see how constrained systems can be treated in terms of superspace integrals. The Berezin integral paves the way to understanding the Batalin–Vilkovisky (BV) formalism [1, 2] in QFT. We will show how this formalism arises, as well as some of its simplest applications. We conclude by discussing the gauge systems in the BV language, the geometrical meaning of the Faddeev–Popov ghosts and show how the BV formalism opens the way for generalizing the notion of symmetrical systems.

2. Motivations for and definition of the Berezin integral

2.1. *Geometry, algebra and supergeometry — an outline*

Suppose we have a manifold. Consider the space of smooth, or analytic, or algebraic real (or complex)-valued functions on it. This space forms a commutative ring, where for any point P and functions f_1, f_2 we set

$$(f_1 \cdot f_2)(P) = f_1(P) \cdot f_2(P)\,. \tag{1}$$

We consider an ideal in this ring formed by functions which vanish on a given subspace. These ideals are partially ordered with respect to inclusion — the smaller the subspace, the bigger the ideal. In particular, the maximal ideals correspond to the smallest subspaces

— points.

Given a ring R we may try to form a space such that functions on this space form the original ring. Let us think that the "points" are maximal ideals I_P. The cosets R/I_P are isomorphic to fields (usually, real or complex numbers or some Galois fields), so the "value" of the element of the ring at a "point" is its image in R/I_P.

A crazy idea that proved to be rather fruitful is to generalize the notion of the manifold by relaxing conditions on the algebra of its functions. In particular, we can consider any commutative ring R or any algebra over any field — the corresponding "manifolds" are schemes from algebraic geometry.

We can replace R by any graded commutative ring — in this case we get the so-called superschemes. One can even completely forget the commutativity condition keeping only associativity — this is the scope of the so-called noncommutative geometry.

Here we will mostly focus on supermanifolds corresponding to $\mathbb{Z}_2$ graded commutative rings (GCR). Such rings can be decomposed as

$$GCR = EVEN \oplus ODD\,. \tag{2}$$

Odd elements mutually anticommute while the rest commute.

The simplest supermanifolds are n-dimensional odd spaces. Their rings are generated by even constants and n odd elements ψ^i. One can show that this space has odd "lines," "planes," etc. but only one point.

2.2. *A warm-up exercise: an integral over circle*

From the standard courses on analysis we know that in order to take an integral over a function F over the interval I we can split the interval in a large number of small subintervals I_i on which the function F is approximately constant, $F \approx F_i$, and then sum up these values multiplied by the lengths of the corresponding intervals,

$$\int_I F\,dx \approx \sum_i F_i\,l_i\,. \tag{3}$$

However, such a definition is totally useless for superspace — odd directions have no points — therefore, no intervals. It means that we should try something else: we must find an appropriate property of the integral and consider it as a definition, provided that this property uniquely determines the value of the integral. Let us try, for instance, the famous relation

$$\int_a^b \frac{df}{dx}\, dx = f(b) - f(a)\,. \tag{4}$$

How does it help? Odd directions have no points, and hence no boundary. Therefore, we should consider functions on the circle, i.e. such that $f(a) = f(b)$. Then

$$\int_{S_1} \frac{df}{dx}\, dx = 0\,. \tag{5}$$

It turns out that this relation determines integral over the circle up to a multiplicative constant. Let x be a coordinate on the circle, $x + 2\pi \sim x$, so we can represent each function on the circle as

$$f(x) = c + \sum_{n>0} \{a_n \cos(nx) + b_n \sin(nx)\}\ . \tag{6}$$

From (5) we conclude that the integral is independent of a_n and b_n! Indeed,

$$\cos(nx) = \frac{d}{dx}\frac{\sin(nx)}{n}\,, \qquad \sin(nx) = \frac{d}{dx}\frac{-\cos(nx)}{n}\,. \tag{7}$$

Hence, the integral we consider is a linear function in c. We just have to fix normalization. On the circle, the standard normalization is $\int 1 \cdot dx = 2\pi$. This completes the theory of integral over the circle.

Note, that here it is important that we choose the vector field $\partial/\partial X$, rather than $v(X)\partial/\partial X$ with some nonconstant $v(X)$. In the latter case we would get a different answer!

Indeed, in general what we must define is a set of vector fields (first-order differential operators)

$$D_a = v_a^i(X)\frac{\partial}{\partial X^i}$$

such that

$$\int_M \mu\, D_a f = 0\,. \tag{8}$$

Note, that these differential operators form a Lie algebra which can be understood as a Lie algebra of the Lie group of transformations of the space M preserving the measure μ.

2.3. *Definition of the Berezin integral*

The great insight of Felix Berezin was that odd directions may be treated in exactly the same way! The only thing to know is the derivative, and this operation is purely algebraic,

$$f(\psi+\epsilon) = f(\psi) + \varepsilon\,\frac{df}{d\psi}.$$

The expansion in ε stops here since, being odd, there is no ε^2. Higher terms are absent too.

Berezin proposed the following definition (here $\mathcal{D}\psi$ is a measure to be defined):

$$\int_{\text{Berezin}} \mathcal{D}\psi\,\frac{\partial f}{\partial\psi} = 0\,. \tag{9}$$

Consider the space of functions of a single odd variable ψ. This space is just two-dimensional since any function of ψ has the form

$$f(\psi) = c_0 + \psi c_1\,.$$

Since

$$c_0 = \frac{\partial(\psi\, c_0)}{\partial\psi}$$

we conclude that the Berezin integral is independent on c_0. It is proportional to c_1. A nice pick for the normalizing factor is to take the proportionality coefficient 1.

To complete the definition we must define the *multi-dimensional* Berezin integral. This can be done by either generalizing the argument above with the derivative, or demanding the integral over the product of two functions to factorize. Both approaches give the same result. For instance, for two-dimensional odd integral we have

$$\int_{\text{Berezin}} \mathcal{D}\psi_1\,\mathcal{D}\psi_2\,(a+\psi_1\,b_1+\psi_2\,b_2+\psi_1\psi_2\,c) = c\,. \tag{10}$$

It corresponds to the Abelian algebra of measure-preserving operators generated by

$$\frac{\partial}{\partial\psi_1} \quad \text{and} \quad \frac{\partial}{\partial\psi_2}\,.$$

The surprising feature of the Berezin integral that follows from its definition is the fact that

$$\int_{\text{Berezin}} \mathcal{D}(c\psi)\, f = \frac{1}{c} \int_{\text{Berezin}} \mathcal{D}\psi\, f\,. \tag{11}$$

This property contrasts a naive expectation that the symbol

$$\int_{\text{Berezin}} \mathcal{D}\psi$$

can be considered as an integral over the differential form $d\psi$ over some virtual space. Indeed, since $d\,(c\psi) = cd\psi$ we would get then c rather than $\frac{1}{c}$ on the right-hand side of Eq. (11).

Thus, the Berezin integral is *not* an integral over a differential top form, as one might naively and mistakenly guess. We will return to this point in Sect. 3.

3. Differential forms as functions on the superspace and their integration

In Sect. 2 we discussed the integral as a linear operation on functions. There is another notion of integrals in classical mathematics, the integral over a differential form over a manifold.

Consider a space M and the space $Fun(M)$ of real-valued functions on M. We may think of M as of the space $Points(M)$, which is the space of points on M. Moreover, we have a *pairing* which takes a point and a function to the value of the function at the given point,

$$Points(M) \times Fun(M) \to \mathbb{R}; \qquad \langle P, f\rangle = f(P)\,. \tag{12}$$

In this pairing the space of functions is a vector space while the space of points is not. We promote it to the vector space of formal linear combination of points. Then the pairing turns out to be bilinear pairing between the vector spaces.

We can consider points as zero-dimensional submanifolds, and look for possible pairings with oriented submanifolds of higher dimension. It turns out that differential forms generalize functions. For example, let us construct an object dual to one-dimensional manifolds and additive with respect to cutting this manifold into pieces. Such object is completely determined by its value on a small interval attached to the point P, i.e. by its value on the tangent vector at the point P.

In fact, this is the definition (up to a smoothness condition) of a differential 1-form. We denote the space of differential 1-forms by $\Omega^1(M)$. By the same token, functions on the surface elements can be shown to be functions on antisymmetrized pairs of tangent vectors at the given point, i.e., 2-forms which span the space denoted $\Omega^2(M)$, and so on.

In this approach, differential forms come equipped with the notion of integral. Let k-chains be the elements of the vector space of formal linear combinations of k-dimensional oriented submanifolds (-1 times a manifold means the manifold with the opposite orientation). There is a pairing

$$\text{(The space of } k\text{-chains)} \times \Omega^k(M) \to \mathbb{R}; \quad \langle C, \omega \rangle = \int_C \omega \,. \tag{13}$$

Note that for k-chains the notion of the boundary ∂ is defined. The boundary of a given k-chain is the $k-1$-chain which is the sum of the boundaries of each summand constituting the chain. Even if we start from a chain which is a single submanifold, its boundary can be a linear combination of submanifolds, i.e., a chain. The simplest example is the oriented interval whose boundary is an end point minus the starting point of the interval.

Now, we can *define* the de Rham exterior differential d as an operation on the differential forms dual to the operator ∂ of taking the boundary,

$$\int_{\partial C} \omega = \int_C d\omega \,. \tag{14}$$

From this point of view d is something that has to be computed from the relation (14).[b]

People advocating this point of view stress the geometrical nature of the integration procedure, while Berezin's approach seems to be algebraic. It seems clear that it is impossible to apply the geometric approach to superspaces since the very notion of superspace is not geometrical.

However, the geometric approach can be easily expressed in terms of the Berezin integral. Indeed, consider the superspace $\Pi(T)M$, which is the tangent bundle to the space M with the reversed parity of the fibers. In simple terms, the tangent bundle TM can be coordinatized by coordinates X^i_α on the base M and coordinates t^i_α along the tangent plane corresponding to the patch U_α on M. On the intersection of patches we have

$$X^i_\alpha = f^i_{\alpha\beta}(X_\beta)\,; \qquad t^i_\alpha = \frac{\partial f^i_{\alpha\beta}}{\partial X^j_\beta}\, t^j_\beta\,. \tag{15}$$

Functions on the space TM correspond just to a set of functions $F_\alpha(X_\alpha, t_\alpha)$ from U_α to $\mathbb{R}$ that are invariant under the transformations (15),

$$F_\alpha(X_\alpha(X_\beta), t_\alpha(X_\beta, t_\beta)) = F_\beta(X_\beta, t_\beta)\,. \tag{16}$$

Expanding in t we see that such functions are symmetric covariant tensors.

The space $\Pi(T)M$ also corresponds to the manifold M covered with the patches and local coordinates X_α and ψ_α. The coordinate transformations are the same,

$$X^i_\alpha = f^i_{\alpha\beta}(X_\beta); \qquad \psi^i_\alpha = \frac{\partial f^i_{\alpha\beta}}{\partial X^j_\beta}\, \psi^j_\beta\,, \tag{17}$$

[b]It is possible, however, to reverse the logic and give an alternative definition of the integral provided the space of differential forms and the exterior differential d are *given*. In this case, the integral over a compact k-dimensional manifold is defined (up to a constant factor) as a linear map from the space of k-forms to reals, vanishing on the image of d.

but now ψ's anticommute. Therefore, the functions on $\Pi(T)M$ correspond to covariant antisymmetric tensors, i.e., differential forms.

From our discussion of the geometrical integral we may anticipate the canonical measure on $\Pi(T)M$. It should correspond to the measure defined on each patch which is invariant under the transformations (17). Such a measure exists indeed and is sometimes called the *canonical Berezin measure.* It has the form

$$\mu_{\text{Ber, can}} = \prod_i dX^i_\alpha \prod_i \mathcal{D}\psi^i_\alpha . \tag{18}$$

Note, that the first (bosonic) factor in (18) is constructed with the help of the local top differential form, while the second has its distinct meaning, defined above. It is not the superdifferential form.

In order to illustrate how things work consider a one-dimensional case. Then the measure is

$$dX_\alpha \mathcal{D}\psi_\alpha ,$$

and under the transformations $X_\alpha = cX_\beta, \; \psi_\alpha = c\psi_\beta$

$$dX_\alpha = cdX_\beta , \quad \text{while} \quad \mathcal{D}\psi_\alpha = \frac{1}{c}\mathcal{D}\psi_\alpha . \tag{19}$$

What about the de Rham operator? When we consider differential forms as functions, the de Rham operator becomes just the vector field. Ona can show that this vector field preserves the canonical Berezin measure.

4. The Hodge star operation as an odd Fourier transform

The simplest application of the Berezin integral is the interpretation of the Hodge *-operation. In particular, we have to interpret the property

$$*^2 = (-1)^{\deg} ,$$

where deg is the degree of the differential form. The Fourier transformation has the same property, and it turns out that the

Hodge star is in fact an odd Fourier transformation. Namely,

$$\tilde{f}(X,\tilde{\psi}) = C(X) \int \mathcal{D}\psi_1 \dots \mathcal{D}\psi_n \exp\left(g_{ij}(X)\tilde{\psi}^i\psi^j\right) f(X,\psi)\,. \tag{20}$$

Note, that if we apply this transformation twice, we will get

$$\begin{aligned}\tilde{\tilde{f}}(\tilde{\tilde{\psi}}) &= C^2 \int \mathcal{D}\psi_1 \dots \mathcal{D}\psi_n \\ &\quad \times \mathcal{D}\tilde{\psi}_1 \dots \mathcal{D}\tilde{\psi}_n \exp\left(g_{ij}\tilde{\psi}^i\left(\psi^j + \tilde{\tilde{\psi}}^j\right)\right) f(\psi) \\ &= C^2 \det g_{ij}\, f(-\tilde{\tilde{\psi}}) = C^2 \det g_{ij}\, (-1)^{\deg} f(\tilde{\tilde{\psi}})\,.\end{aligned} \tag{21}$$

That is why it is natural to take

$$C = (\det g_{ij})^{-\frac{1}{2}}\,.$$

5. The Mathai–Quillen representative and supermanifolds as fractions

The space of fields in QFT may be modeled by a manifold. There are two classical ways of constructing new manifolds from the old ones: to take the coset with respect to the action of a group (which will be referred to as the gauge group), and consider zeroes of a function; the procedure is called writing a (system of) *equation*(s) (in mathematics) or imposing a (set of) *constraint*(s) (in physics). The question is how to describe an integral over the new manifold in terms of the integral over the old one. We will return to cosets after we discuss the Batalin–Vilkovisky formalism, while constraints can be considered right here. We will study the simplest possible example, the integral over the zeroes of the function f,

$$\int_{C\in M} \omega = \int_M \delta_C \omega\,, \qquad C = \{p \mid f(p) = 0\}\,. \tag{22}$$

We will present and interpret the integral representation for the δ function in terms of the supermanifolds and the Berezin integral (in the physical literature see Ref. [3]).

The first approximation is to consider a regularized δ function

$$\delta^{\text{naive},m} = \exp\left(-m^2 f^2\right) m\,.$$

Indeed, this function is concentrated along the zeros of f and its integral is independent of m as $m \longrightarrow \infty$. However, it would be an incorrect answer since δ has to be a 1-form rather than a function (in order to match dimensions of the forms that have to be integrated over C and M). Moreover, the integral of $\delta^{\text{naive},m}$ changes as $f \longrightarrow c\,f$ while the set of zeroes does not. These two problems are cured by setting

$$\delta^m = \frac{1}{\sqrt{\pi}} \exp\left(-m^2 f^2\right) m\, df\,. \tag{23}$$

We start from uplifting mdf to the exponent. Namely, we introduce an odd variable λ' such that

$$\delta^m = \frac{1}{\sqrt{\pi}} \int \mathcal{D}\lambda' \exp\left(-m^2 f^2 + \lambda' m df\right)\,. \tag{24}$$

The above expression is correct; however, its geometric meaning is unclear. The terms in the exponent look very different, while purely odd integral is a bit strange and the large-m limit is hard to take.

This can be cured by introducing an auxiliary even variable l', a Lagrange multiplier, so that we arrive at

$$\delta^m = \frac{1}{2\pi} \int \mathcal{D}\lambda'\, dl' \exp\left(il'mf + \lambda' m df - \frac{l'^2}{4}\right)\,. \tag{25}$$

Introducing new variables l and λ,

$$2\pi m l = m l'\,, \quad \lambda = m\lambda'$$

we get the final expression

$$\delta^m = \int \mathcal{D}\lambda\, dl \exp\left(2\pi i l f + \lambda df - \frac{(l\pi)^2}{m^2}\right)\,. \tag{26}$$

Now we have the integral over the canonical Berezin measure. Moreover, the expression in the exponent has a naive limit as $m \to \infty$. We will see that this limit has a nice and surprising geometric meaning. If we interpret df as a function on the superspace $\Pi(T)M$, with the coordinates X and ψ, then expression in the exponent has the form

$$d^{\rm super}(f\lambda) = \left(\psi^j \frac{\partial}{\partial X^j} + 2\pi i l \frac{\partial}{\partial \lambda}\right)(f\lambda)\,. \tag{27}$$

The new operator $d^{\rm super}$ may be considered as a de Rham operator on a superspace $M^{\rm super}$ with the even coordinates X^i and odd coordinate λ. All this allows us to rewrite the original integral as follows:

$$\int_{C\in M} \omega = \int_{\Pi(T)M^{\rm super}} \mu_{\rm Ber}\left((\Pi(T)M^{\rm super}\right)\omega \exp\left((-d^{\rm super}(f\lambda)\right)\,. \tag{28}$$

In particular, we can consider supermanifold as something like a *fraction* of the space of variables over the space of equations. The integral (28) is a way to simplify this fraction with the help of the function f. If ω is closed with respect to $d^{\rm super}$, one can show that the result of integration does not depend on the particular choice of f. This can be understood as a manifestation of the fact that different ways to simplify a given "fraction" are equivalent (when applied to "fraction manifolds" with the insertion of a closed form)!

This construction is widely used in topological quantum field theories. In particular, in type A topological sigma models, the space M is an infinite-dimensional space of all maps from the world sheet to the target space, f is replaced by the infinite-dimensional space of holomorphicity equations, the cycle C stands for holomorphic maps — instantons. The ψ and λ fermions are just twisted fermions of the standard supersymmetric sigma model, $d^{\rm super}$ is what is called scalar nilpotent Q symmetry, and the parameter m is often referred to as the $\bar{t}$ coupling constant.

6. The BV master equation as an odd Fourier transform of the de Rham equation

As we will see shortly, the Batalin–Vilkovisky formalism — the most powerful formalism for treating symmetries in (functional) integrals — can be understood in terms of the odd Fourier transformation for the theory of the de Rham cohomology [4] (after Georges de Rham).

Recall that this cohomology, intended for studying integrals over the closed submanifolds up to their deformations, is defined as the quotient space of closed forms ω modulo the image of the exterior differential,

$$d\omega = 0, \qquad \omega \sim \omega + d\nu\,. \tag{29}$$

One of the most important operations on the de Rham cohomologies is the procedure of integration over the compact fiber. Namely, consider the manifold M which is a family of compact fibers F parametrized by the base B. Given a closed form Ω on M one produces a closed form on the base B by integrating it over the compact fiber,

$$\Omega_B = \int_F \Omega\,. \tag{30}$$

One can check that the integral over the exact forms produces exact forms. This follows from the decomposition of the total differential as a sum of differentials over the fiber and the base,

$$d_M = d_F + d_B\,, \tag{31}$$

and

$$\int_F d_F \nu = 0\,, \tag{32}$$

due to the compactness of the fiber. It means that we actually have a map from the cohomology classes of the total space to the cohomology classes of the base.

The BV formalism is an equation on the $\Pi(T^*)M$ which is a cotangent bundle to M with reversed parity of the fibers. It provides equations on functions in X^i and X_i^*, where X_i^* is the odd variable which is the coordinate not on the tangent but rather on the cotangent bundle with reversed parity.

Such functions naturally appear if we perform the odd Fourier transformation over the fiber (as in the theory of the Hodge duality). Now we have no metric to identify the tangent and cotangent directions. Moreover, in performing this Fourier transformation we need a measure on the fiber, which will be denoted as Ω,

$$g(X^*, X) = \int \Omega(X)\, \mathcal{D}\psi^1 \ldots \mathcal{D}\psi^n\, f(\psi, X)\,. \tag{33}$$

Under this transformation the de Rham operator considered as a first-order differential operator on the superspace transforms into the following second-order operator:

$$\Delta = \frac{\partial}{\partial X^i} \frac{\partial}{\partial X^*_i}\,. \tag{34}$$

The "closedness" conditions turn into pre-BV equations

$$\Delta g(X^*, X) = 0\,. \tag{35}$$

The operation of integration over the fiber becomes the integration over the Lagrangian submanifold in the fiber corresponding to $X^* = 0$.

The second step in the BV formalism is to consider special functions g of the form

$$g(X^*, X) = \exp\left(\frac{1}{\hbar} S(X^*, X, \hbar)\right), \tag{36}$$

where S is regular at $\hbar = 0$.

Equation (35) leads to the equation for S which is referred to as the *master* equation,

$$\{S, S\}_{\mathrm{BV}} + \hbar \Delta S = 0\,, \tag{37}$$

where the BV bracket is defined as

$$\begin{aligned} \{g_1, g_2\}_{\mathrm{BV}} &= \Delta\,(g_1 g_2) - g_1 \Delta g_2 - (\Delta g_1)\, g_2 \\ &= \frac{\partial g_1}{\partial X^i} \frac{\partial g_2}{\partial X^*_i} - (1 \leftrightarrow 2)\,. \end{aligned} \tag{38}$$

The BV-bracket was known in mathematics as the Schouten bracket. It generalizes the well-known Lie bracket (the commutator of the vector fields) to the polyvector fields (the polyvector fields are nothing but the contravariant antisymmetric tensors). In particular,

$$\{f_1, f_2\}_{BV} = 0; \quad \{v^i X^*_i, f\}_{\rm BV} = v^i \partial_i f,$$

$$\{v^i X^*_i, u^j X^*_j\}_{\rm BV} = (v^i \partial_i u^j - u^i \partial_i v^j) X^*_j\,, \tag{39}$$

$$\{\pi^{ij} X^*_i X^*_j,\, \pi^{pq} X^*_p X^*_q\}_{\rm BV} = 4\, \pi^{ij} \frac{\partial \pi^{pq}}{\partial X^i} X^*_j X^*_p X^*_q\,. \tag{40}$$

7. An exercise in the BV language

Evaluating the master BV equation (37) at $\hbar = 0$ we get the so-called classical BV equation,

$$\{S(X^*, X, 0), S(X^*, X, 0)\}_{BV} = 0\,. \tag{41}$$

In particular,

$$S = \pi^{ij} X^*_i X^*_j$$

solves the classical BV equation if the bivector π^{ij} defines a Poisson bracket. Indeed, if we study the Jacobi equation for the bracket given by the bivector we will obtain the right-hand side of (40).

It is known that for invertible π the Poisson equation is equivalent to the "closedness" of the 2-form,

$$\omega_{ij} = (\pi^{-1})_{ij}\,.$$

This well-known fact always seemed to be a miracle since it connects solutions to the system of quadratic equations with solutions to the linear equation!

However, this miracle has a natural explanation from the BV-integral point of view. Indeed, consider an obvious closed form constructed from the closed 2-form ω

$$d\omega = 0 \Rightarrow d \exp\left(\frac{1}{\hbar}\,\omega\right) = 0\,. \tag{42}$$

Now, let us perform an odd Fourier transformation

$$\int \Omega(X)\mathcal{D}\psi^1 \dots \mathcal{D}\psi^n \exp\left(\frac{1}{\hbar}(\omega_{ij}\psi^i\psi^j + X_i^*\psi^i)\right)$$

$$= \exp\left(\frac{1}{\hbar}S_\omega(\hbar)\right) = \exp\left(\frac{1}{\hbar}(\pi^{ij}X_i^*X_j^* + O(\hbar))\right), \tag{43}$$

where the $O(\hbar)$ correction is the determinant of the matrix ω_{ij}. As we already explained, $S_\omega(\hbar)$ solves the BV equation. Thus, $S_\omega(0)$ solves the classical BV equation. π is the Poisson bivector field. Miracle explained.

8. The BV integral

The main idea of the BV approach is that the only integrals are the *BV integrals*, i.e. those over the exponent of the BV action over Lagrangian manifolds (a d-dimensional subspace of the $2d$-dimensional BV spaces such that the so-called BV symplectic form

$$\delta X^i \wedge \delta X_i^*$$

vanishes on it). The *only property* of the BV integrals is that the result of the BV integration of

$$\exp\left(\frac{1}{\hbar}S(X^*, X, Y^*, Y, \hbar)\right)$$

gives effective actions which satisfy the master equation (37) on the space of parameters Y, Y^* [1, 2].

The simplest example is the action $f(X, Y)$ (which does not depend on the antifields X^* and Y^*). We consider the coordinates Y, Y^* as parameters, and $X^* = 0$ as a Lagrangian submanifold. Clearly, the integral is independent of Y^* and therefore solves the master equations.

The second example comes from rethinking of the Berezin integral which is an odd Fourier transform relating the symplectic and Poisson structures. To this end we introduce the BV space with the coordinates

$$X^i, X_i^*, \ \psi^i, \ \psi_i^*$$

with ψ_i^* being the antifield to the field ψ^i. Hence, ψ_i^* has the opposite parity and is thus even. Then the expression in the exponent in Eq. (43) is just another example of the action solving the BV equations.

Indeed, the term $\psi^i X_i^*$ in the BV action corresponds to the vector field $\psi^i \partial/\partial X_i$ which is just a de Rham exterior differential d. In the BV language, the phrase "the differential 2-form ω is close" simply means that the function $\omega_{ij}\psi^i\psi^j$ is d-invariant.

Now we will interpret the Fourier transformation as a BV integral. Let us consider X, X^* as parameters, ψ, ψ^* as coordinates on the BV space, and $\psi^* = 0$ as the equation of a Lagrangian submanifold. Due to the only propety of the BV integral (mentioned above) the effective action $\pi^{ij} X_i^* X_j^*$ solves the BV equations, i.e., it is the Poisson bivector field.

9. The BV formalism, symmetries and gauge fixing

In gauge theories we start from the space of fields with the action of a symmetry group. We consider actions which are the functional of fields invariant under the action of a symmetry group and perform integration over the space of orbits. In the infinite-dimensional case, the standard way to calculate this integral is to pick up representatives on each orbit (we say that they satisfy the gauge fixing condition) and integrate with a very special measure over the space of representatives. A convenient way to study such measure is to calculate the Berezin integral over special odd fields which are usually referred to as *ghosts*.

As we will explain shortly, the BV formalism is perfectly well suited for this. We will see the geometrical meaning of the ghosts even before the gauge fixing.

9.1. *Symmetric systems in the BV language*

Here we will consider the space of QFT fields as some (super)space M. Certain fields will be coordinates X^i on this space, or functions of these coordinates. The classical system is determined by the action $f(X)$ which is a function on M. The classical symmetries are described by vector fields $v_a^i(X)\frac{\partial}{\partial X^i}$. The subscript a labels these fields. We say that a given action f has a (super)Lie algebra with the structure constants C_{ab}^c as its symmetry algebra if

$$v_a^i(X)\frac{\partial f}{\partial X^i} = 0\,, \tag{44}$$

and

$$\left[v_a^i(X)\frac{\partial}{\partial X^i}, v_b^j(X)\frac{\partial}{\partial X^j}\right]_\pm = C_{ab}^c v_c^i(X)\frac{\partial}{\partial X^i}\,, \tag{45}$$

where $[,]_\pm$ stands for the graded commutator. In the BV language the system of equations (44) and (45) has the following interpretation. Consider the superspace

$$\begin{aligned} M_{\mathrm{BV}} &= \Pi(T^*)M \times \mathcal{G}^* \times \Pi(\mathcal{G}) \\ &= \Pi(T^*)\big(M \times \Pi(\mathcal{G})\big) \end{aligned} \tag{46}$$

with the coordinates

$$X^i,\ X_i^*,\ c_a^*,\ c^a\,.$$

Here the parity of the so-called antifield X_i^* is opposite to that of X^i. The space $\Pi(\mathcal{G})$ is the (super)Lie algebra with inverted parity. In particular, for ordinary (even) symmetries, the coordinates on $\Pi(\mathcal{G})$ are odd fields c^a. They are involved in the parametrization of the general symmetry transformation as $c^a v_a^i \partial/\partial X^i$. We will see that these c^a's (originally introduced in the 1950s in the cohomology theory of the Lie algebras) would turn out to be *c-ghosts* (which form a half of the Faddeev–Popov ghosts).

Note, that c_a^* is a coordinate on $\mathcal{G}^*$ which is even for ordinary symmetries; they are antifields to ghosts c, but they are not what people ordinarily call the Faddeev–Popov antighost (we will reserve the special name *b-ghosts* for the latter; we will encounter b-ghosts below).

The system of equations (44) and (45) is equivalent to the statement that

$$S(X, X^*, c, c^*) = f(X) + c^a v_a^i X_i^* + \frac{1}{2} C_{ab}^c c^a c^b c_c^* \tag{47}$$

solves the classical BV equation.

9.2. *Gauge fixing as the BV integral*

Suppose we have a symmetric system and we want to fix a gauge. In the BV language it means that we choose a very special Lagrangian manifold, such that the variables X are restricted by the constraint

$$f_\alpha(X) = 0; \quad \alpha = 1, \ldots, \dim \mathcal{G}\,. \tag{48}$$

Simultaneously we have to change the $X^* = 0$ condition into

$$X_i^* = \frac{\partial f_\alpha}{\partial X^i}\, b^\alpha\,. \tag{49}$$

Here b^α are the coordinates on the Lagrangian submanifold (they have the same parity as X^*, in particular, they are odd when the coordinates X are even). Let us check that Eq. (49) actually represents a Lagrangian submanifold,

$$\delta X^i \wedge \delta X_i^* = \delta X^i \frac{\partial f_\alpha}{\partial X^i} \delta b^\alpha + \delta X^i \wedge \delta X^j \frac{\partial^2 f_\alpha}{\partial X^i\, \partial X^j} = 0\,, \tag{50}$$

where the first term vanishes due to the gauge fixing condition (48), and the second one vanishes due to the contraction between antisymmetric and symmetric tensors.

In the ghost sector we choose the Lagrangian submanifold

$$c^* = 0\,.$$

Realizing the gauge fixing condition (48) with the help of the Lagrange multipliers β^α we can rewrite the integral in the symmetric system as

$$\int \Pi_\alpha\, Db^\alpha\, d\beta^\alpha\, \Pi_a\, Dc^a\, \Omega(X)$$

$$\times \exp\left(f(X) + c^a v_a^i \frac{\partial f_\alpha}{\partial X^i} b^\alpha + 2\pi\, i \beta^\alpha f_\alpha \right), \tag{51}$$

where $\Pi_a\, Dc^a$ is the canonical measure on the Lie algebra with the inversed parity. In Eq. (51) we can easily recognize the standard

Faddeev–Popov formula, where the coordinates b^α become the Faddeev–Popov b-ghosts.

10. Generalizations of classical symmetries

10.1. *Differential graded Lie algebras and homotopical Lie algebras*

Expression (47) is rather inspiring for generalization of the classical notion of symmetry in the framework of the BV formalism. Let us start from the simplest space M, a point. Then we observe that the only solution to the BV equation is $C^c_{ab}c^a c^b c^*_c$. This solution is linear in antighosts but quadratic in ghosts. We can ask what will happen if we also consider terms linear in ghosts. Naively, this cannot happen since the corresponding term seems to be odd. However, if we extend the Lie algebras to super-Lie algebras this is no longer a problem. As a result, we can try

$$S(c, c^*) = \left\{ q^e_a c^a + \frac{1}{2} C^e_{ab} c^a c^b \right\} c^*_e . \tag{52}$$

The first term determines a linear operator q on the Lie algebra. The BV equation means that q is a differential, i.e., $q^2 = 0$, and q differentiates the bracket, i.e., there is a Leibniz rule,

$$q([A, B]) = [q(A), B] + (-1)^{\mathrm{par}\, A}[A, q(B)] . \tag{53}$$

Such a structure is known as the *differential graded Lie (DGL) algebra.*

We may also include higher terms in c by adding terms

$$C^e_{a_1 \ldots a_k} c^{a_1} \ldots c^{a_k} c^*_e$$

in Eq. (52). Then the BV equations require $q^2 = 0$, and the Leibniz rule and homotopy rules, with the simplest expression

$$C^b_{[a_1 a_2} C^f_{b a_3]} = q^f_d C^d_{a_1 a_2 a_3} - C^f_{[a_1 a_2 b} q^b_{a_3]} . \tag{54}$$

Here the square brackets stand for antisymmetrization (which must include additional signs due to parity of the variables). The left-hand side of Eq. (54) is the Jacobi identity, while the right-hand side can be interpreted as a commutator between the differential q and a new $3 \to 1$ operation $C^d_{a_1 a_2 a_3}$. Thus, Eq. (54) means that the "Jacobi identity" considered as the $3 \to 1$ operation is q-exact. Similar relations exist for higher operations. Such an algebra is called the homotopy Lie algebra or L_∞ algebra.

10.2. *Algebroids*

Another generalization of the notion of symmetry takes place when what was previously called structure constants start to depend on X, namely when we consider a solution to BV equations of the form

$$S(X, X^*, c, c^*) = f(X) + c^a v^i_a X^*_i + (1/2) C^c_{ab}(X) c^a c^b c^*_c \,. \tag{55}$$

Such structure is referred to as the *Lie algebroid* [5]. It may look rather weird but mathematical physicists had already encountered it — believe it or not — in supersymmetric gauge theories in the Wess–Zumino gauge! In these theories the anticommutator of two supersymmetries $\{Q_\alpha, Q_\alpha\}$ is a sum of a shift and a gauge transformation with the gauge parameter $\gamma^m A_m$. In the BV language this means that the action (in the Abelian gauge theory, just for simplicity)

$$\begin{aligned} S &= S_{\mathrm{SYM}} + c v_{\mathrm{gauge}} + \epsilon^\alpha v_{\mathrm{super},\alpha} + \eta^\mu v_{\mathrm{shift},\mu} \\ &\quad + \frac{1}{2}\left(\eta^*_\mu + c^* A_\mu\right)\left(\epsilon \gamma^\mu \epsilon\right) \end{aligned} \tag{56}$$

solves the BV equations. Here the vector fields (on the space of fields) v_{super}, v_{shift} and v_{gauge} correspond to supersymmetry, shift symmetry and gauge symmetry, respectively, while ϵ, η and c are the corresponding ghosts, A_μ is the Abelian gauge field.

10.3. *Higher terms in antifields and a new notion of symmetry*

In all examples above the action was linear in antifields. Now, consider the case where it has quadratic terms. This can happen if the BV action is effective, i.e., it is obtained after taking a BV integral.

Consider the simplest example: the space $\mathbb{R}^n$, on which $SO(n)$ with the generators t_{ab} acts, and the invariant function

$$f = -X_1^2 - \ldots - X_n^2 .$$

The BV action takes the following form:

$$S = -\sum_{i=1}^{n} X_i^2 + \sum_{i,j=1}^{n} c_{ij} X_i^* X_j + \sum_{i,j,k=1}^{n} \frac{1}{2} c_{ik}^* c_{ij} c_{jk} . \tag{57}$$

Now, we integrate S over X_n fixing $X_n^* = 0$ and get

$$S_{\text{eff}} = -\sum_{i=1}^{n-1} X_i^2 + \sum_{i,j=1}^{n-1} c_{ij} X_i^* X_j + \sum_{i,j,k=1}^{n} \frac{1}{2} c_{ik}^* c_{ij} c_{jk}$$

$$- \frac{1}{4} \sum_{i,j=1}^{n-1} c_{in} c_{jn} X_i^* X_j^* . \tag{58}$$

Without the last term the action (58) would imply something non-existing — a nontrivial linear action of $\mathfrak{o}(n)$ on $\mathbb{R}^{n-1}$. However, the last term saves the game. Surprisingly, it generalizes the notion of systems with symmetries.

Indeed, now the symmetry of a system is not only an invariant function and a representation of the Lie algebra of symmetries by vector fields. The new element is a map from $\Lambda^2(\mathfrak{o}(n))$ to the space of *bivectors*,

$$\pi_{kn,ln}^{ij} = \frac{1}{4} \left(\delta_k^i \delta_l^j - \delta_k^j \delta_l^i \right) , \tag{59}$$

such that the vector fields, bivectors, and the symmetric function are mixed up in a single relation [8] (which we will present in the general case),

$$\{\pi_{ab}, f\} + \{v_a, v_b\} = C_{ab}^c v_c . \tag{60}$$

Once again, one may suspect that relations such as (60) are kind of exotic. However, we do meet them in the description of supersymmetric systems (see e.g. [6, 7]). They appear when we integrate out auxiliary fields, as we did in the description of the regularized delta-form. Indeed, the de Rham operator can be considered as a symmetry that squares to zero (with its ghost called ϵ). Therefore, before integrating out the p-fields we have

$$S = 2\pi i l f(X) + \lambda f'(X)\psi - 2\pi^2 l^2 + \epsilon(X^*\psi + 2\pi i \lambda^* l)\,. \qquad (61)$$

Integrating I out we then get

$$S = -\frac{1}{2} f^2 + \lambda\, f'(X)\,\psi + \epsilon\,(X^*\psi - \lambda^* f) - \frac{1}{2}\,\epsilon^2(\lambda^*)^2\,. \qquad (62)$$

The last term is a new bivector. Similar things happen in all supersymmetric theories.

11. Instead of conclusion: Dreams on a BV M-theory

M-theory in modern mathematical physics stands for a hypothetical supersymmetric theory with a complicated space of vacua. Near special points in the space of vacua the theory has distinct reincarnations: either as one of several superstring theories or as eleven-dimensional supergravity with a membrane. We would like to understand M-theory as a phenomenon in a certain unified theory: when the unified theory is defined over a complicated moduli space but it looks more familiar on different boundaries. In the example described above we know no detailed description of the unified theory while we understand, to some extent, its various limits.

Looking at the BV actions we observe a similar pattern. At a generic point the BV action is just some function of even and odd fields and antifields, solving the BV equations. These solutions just have no standard geometrical meaning.

However, as we approach the regions where the BV action degenerates (in particular, when it has no linear term), we may assume the linear structure on the space of fields. Then we can try to understand the limiting solution to the BV equation geometrically. Above we have mentioned several possible patterns: constrains, system with symmetries, and so on.

Moreover, the same solution to the BV equation may have different degeneracies, and therefore different geometrical interpretations. In this case, we say that they are in *duality relation* to each other.

Concluding, I would like to emphasize that all these perspectives became possible due to the development of supergeometry. Felix Berezin gave birth to supergeometry and supercalculus, with far reaching consequences for mathematics and physics, with a long-lasting impact.

Acknowledgements

This work was supported in part by grant RFBR 07-02-01161, by INTAS-03-51-6346, NWO project 047.011.2004.026 and NSh-8065.2006.2.

Appendix: Books by F.A. Berezin

- Метод вторичного квантования, Москва, Наука, 1965
- Ф. А. Березин, М. А. Шубин, Лекции по квантовой механике, Москва, Издательство МГУ, 1972
- Лекции по статистической физике, Москва, Издательство МГУ, 1972
- Лекции по статистической физике, 2-е издание, Ижевск, Институт компьютерных исследований, 2002
- Метод вторичного квантования, Москва, Наука, 2-е дополненное издание, 1986
- Метод вторичного квантования (Шедевры мировой физико-математической литературы), Москва, Издательство ИО НФМИ, 2000
- Ф. А. Березин, М. А. Шубин, Уравнение Шредингера, Москва, Издательство МГУ, 1983
- Введение в алгебру и анализ с антикоммутирующими переменными, Москва, Издательство МГУ, 1983

• Method of Second Quantization, New York, Academic Press, 1966

• Lie supergroups, Moscow, ITEP, 1977

• Introduction to Superanalysis, Dordrecht, D. Reidel Publishing Company, 1987

• Introduction to Superanalysis, Berlin, Springer, 2001

• F. A. Berezin, M.A. Shubin, The Schrödinger Equation, Berlin, Springer, 2002

• F. A. Berezin, Lectures on Statistical Physics, Singapore, World Scientific, 2007

References

1. I. A. Batalin and G. A. Vilkovisky, *Gauge Algebra And Quantization,* Phys. Lett. B **102**, 27 (1981); *Quantization Of Gauge Theories With Linearly Dependent Generators,* Phys. Rev. D **28**, 2567 (1983), (E) D **30**, 508 (1984); I. A. Batalin and E. S. Fradkin, *A Generalized Canonical Formalism And Quantization Of Reducible Gauge Theories,* Phys. Lett. B **122**, 157 (1983); B. L. Voronov and I. V. Tyutin, *Formulation Of Gauge Theories Of General Form. I,* Theor. Math. Phys. **50** (1982) 218.
2. A. S. Schwarz, *Geometry of Batalin–Vilkovisky quantization,* Commun. Math. Phys. **155**, 249 (1993) [hep-th/9205088]; *Semiclassical Approximation in Batalin–Vilkovisky Formalism,* Commun. Math. Phys. **158** (1993) 373 [hep-th/9210115].
3. S. Cordes, G. Moore, S. Ramgoolam, *Lectures on Two-Dimensional Yang–Mills Theory, Equivariant Cohomology and Topological Field Theories,* Nucl. Phys. Proc. Suppl. **41**, 184 (1995) [hep-th/9411210].
4. R. Bott, L.W. Tu, *Differential Forms in Algebraic Topology,* (Springer-Verlag, Berlin, 1995).
5. K.Mackenzie, *Lie Groupoids and Lie Algebroids in Differential Geometry,* (Cambridge University Press, 1987).
6. M. B. Green and C. M. Hull, *Quantum Mechanics of a Twisted Superparticle,* Nucl. Phys. B **344**, 115 (1990); *The Quantum Mechanics of an $n+1$ Superparticle in an Extended Superspace,* Mod. Phys. Lett. A **5**, 1399 (1990); E. Bergshoeff, R. Kallosh and A. Van Proeyen, *Superparticle Actions And Gauge Fixings,* Class. Quant. Grav. **9**, 321 (1992).
7. V. Aleksandrov, D. Krotov, A. Losev, V. Lysov, *A Pure Spinor Superfield Formalism,* in preparation.
8. D. Krotov, A. Losev, V. Lysov, to appear.

FELIX ALEKSANDROVICH BEREZIN *

A Brief Scientific Biography †

ROBERT MINLOS

Dobrushin Mathematics Laboratory
Institute for Information Transmission Problems
Ermolovoi 19, 101447 Moscow, Russia
and
Independent University of Moscow, Bolshoy Vlasyevskiy Pereulok 11,
119002 Moscow, Russia
minl@iitp.ru

1. What He Achieved in Science

An Overview of F. A. Berezin's Work,
The Modern Understanding of Mathematical Physics

In his relatively brief life (he died in an accident before reaching the age of 50), F. A. Berezin succeeded in doing a great deal in mathematics and mathematical physics. Not only did he leave a deep trace in several branches of mathematics that existed before him (group representation theory, the spectral theory of operators, quantum mechanics, statistical physics, constructive quantum field theory), but he also initiated several new concepts, methods, and theories: a general approach to the quantization problem, the construction of the second quantization formalism in terms of functional

*In the preparation of this essay I used materials made available to me a few years ago by A.A. Kirillov, D.A. Leites, V.N. Sushko, M.A. Shubin. Several facts I learned from N.D. Vvedenskaya. I am grateful to all of them.

†Reprinted from *American Mathematical Society Translations*, Series 2, vol. 175, pp. 1-13, 1996. Also published in Lett. Math. Physics, vol. 74, 5-19, 2005. French translation in La Gazette des Mathématiciens, le bulletin interne de la Société Mathématique de France, 110, 30-44, Octobre 2006.

integrals, which later became the so-called "calculus of symbols" (a forerunner of the theory of pseudo-differential operators), and finally (this was his most important and long nurtured achievement) the theory of supersymmetry and supermanifolds, i.e., what mathematicians now usually call supermathematics.

Further we shall discuss all these topics in more detail. Here I would only like to stress that perhaps the most valuable and important characteristic of Berezin's mathematical life was not his concrete achievements, but the overall stubborn direction of his research, whose main backbone was mathematical physics. He was one of the very few people who transformed mathematical physics into what it has become today. In fact, until the end of the 1950s, the expression "mathematical physics," at least in Russia, was mainly associated with the study of special types of differential equations arising in physics theories (the wave equation, the heat equation, etc.). Berezin was one of the first to notice that, as the old saying goes, the old barrels are not too ancient for the young wine, and the term "mathematical physics" should be applied to a much wider class of mathematical problems, namely to all theories and structures in mathematics that arise in attempts to clearly understand the fundamental physical theories (quantum physics, kinetics, statistical physics, gravitation). Today mathematical physics, precisely in this understanding, has developed tremendously and had attracted many mathematicians (and even some physicists), while some 35-40 years ago, at the very outset of Berezin's scientific life, nearly all physicists regarded this activity with poorly disguised sneers, while the mathematicians did not disguise that they couldn't care less. One needed a great deal of courage and determination, being aware of this total lack of understanding and secretly suffering from it, to persevere in working in the chosen direction.

Thus, in a few large strokes, we can sketch the main inner motivations of the mathematical work of F. A. Berezin.

2. Early Years

Family, School, University, the Graduate Studies that Never Took Place

Alik (Felix Alexandrovich) Berezin was born on April 25, 1931, in Moscow to a typical intellectual family: his father was an economist, his mother was a doctor. Alik's parents separated early, and he was brought up by his mother and her parents. In 1948, after graduating from high school, he entered the first year of the Mechanics and Mathematics Department of Moscow State University.

His interest in mathematics arose long before that: from the 8th grade he began to participate, on a routine basis, in school mathematical Olympiads, very absorbing mathematical problem-solving competitions organized by young enthusiasts (mostly graduate and undergraduate students) every spring at the Mechanics and Mathematics Department. These same enthusiasts conducted weekly mathematical "circles," something like math seminars for beginners, where elegant theorems and even fragments of mathematical theories, accessible to high school students, were presented, difficult problems were discussed and solved on the spot by the participants. Alik Berezin took part in the work of such a circle, headed by E. B. Dynkin, then still a graduate student.

In his first years at the university he also participated in Dynkin's seminar (for undergraduates), which in fact was the continuation of the circle for high school students. This seminar had two main topics (algebra and probability), corresponding to E. B. Dynkin's two main research interests at the time. Berezin was more interested in algebra and received a strong introduction to the subject, which was to serve him well in all his subsequent work. In the math circle, and later in Dynkin's seminar, Berezin made a very early acquaintance with several budding mathematicians, who were then his fellow students at the department (R. Dobrushin, S. Kamennomostskaya, F. Karpelevich, R. Minlos, I. Shapiro-Pyatetski, N. Vvedenskaya, and A. Yushkevich). These acquaintances, many of whom grew to be life-long friends, were to play an essential role in his life. A few years later, while still an undergraduate, Alik Berezin began to participate in the famous Gelfand seminar, and for a long period of time fell under the influence of Izrael Gelfand. In this seminar he wrote his first important research paper on group representation theory (see below).

In 1953 Berezin graduated from the Mechanics and Mathematics Department of Moscow University. Although by that time he had established himself as a talented young research mathematician (he was apparently the strongest student in his graduating class), he was not recommended for graduate work: in the last years of Stalin's life, anti-Semitism had become a state policy, and Berezin, whose mother was Jewish, was automatically denied this privilege (by that time practically all ethnic Jews, could not even become undergraduate students at Moscow University). For three years, Berezin taught mathematics in one of the Moscow high schools, continued to attend Gelfand seminar and to do research in representation theory. In 1956, with the advent of the Khrushchev liberalization, the atmosphere at the Department of Mechanics and Mathematics changed somewhat for the better, so that I. G. Petrovski, then the Rector of the University, succeeded in giving a job to Berezin at the Chair of Theory of Functions and Functional Analysis at the insistence of I. M. Gelfand. Berezin was only 25, and he was to work at that chair until the end of his life.

3. The First Period of Work at the University

The Chair in the 1950s and '60's,
How Mathematical Physics Started,
Works of the First Period, Berezin as a Teacher

The first years of his work occurred in a period of absolutely exceptional intellectual and spiritual revival that characterized the Mechanics and Mathematics Department at the end of the 1950s and the 1960s. This was especially obvious at the chair where Alik worked. Until the mid-1950s the chair, headed for many years by the marvelous and childishly pure D. E. Menshov, had mostly consisted of specialists in the theory of functions of a real or complex variable (D. E. Menshov, N. K. Bari, and A. I. Markushevich). During the subsequent years this group was also augmented by several qualified experts (P. L. Ulyanov, B. V. Shabat, A. G. Vitushkin, A. A. Gonchar, E. P. Dolzhenko and their pupils), but the most intensive development of the chair took place along the lines of functional analysis, the direction headed by I. M. Gelfand. Thus, R. A. Minlos was hired together with Berezin, and shortly afterward G.E. Shilov came followed by his pupil A. G. Kostuchenko. A few years later a large group of I. M. Gelfand's and G. E. Shilov's brilliant pupils were working there (A. A. Kirillov, V. P. Palamodov, E. A. Gorin, and others). At the beginning of the 1960s Professor B. M. Levitan was invited to the chair, and for several years S. V. Fomin and V. M. Tikhomirov worked there.

Thus, thanks to the efforts of I. M. Gelfand and G. E. Shilov and the support of I. G. Petrovski, the chair acquired a first class group of analysts; no other university in the world could boast of a group at the same level. This team, which was occasionally supplemented by good specialists as the years went by, continued to exist with almost the same members until the early 1990s, when the progressive disintegration of the department (which began in the late 1950s under the deanship of P. M. Ogibalov) passed from the hidden phase to the overt one. Of course, the chair suffered several losses during this long period: the death of G. E. Shilov in 1975, Berezin's death in 1980. And in fact I. M. Gelfand lost interest in affairs of the chair in the late 1960s. But in the 1950s and '60s, the intense scientific life at the chair, the appearance of young and talented undergraduate and graduate students, created an exhilarating and beneficial background for research.

In 1957 Berezin defended his *candidate* dissertation, [a] which incorporated his paper on Laplace operators on semi-simple Lie groups [1]. This paper contained the following remarkable result: the descrip-

[a]The academic hierarchy in Russia follows the German rather than the Anglo-American pattern. An approximate Russian equivalent of PhD in the US is the so called *candidate* degree. The highest academic degree, doctoral, is analogous to the German *Habilitation.* The doctoral dissertation is usually presented at a mature stage of the academic career; only a fraction of the *candidate* degree holders make it to the doctoral level. –Editor's note

tion of all irreducible infinite-dimensional representations of complex semi-simple Lie groups in Banach spaces. In modern language Berezin's theorem may be stated as follows: any irreducible representation of the group G is isomorphic to a subfactor of an elementary representation (i.e., a representation induced by a one-dimensional character of a Borel subgroup). The depth of this fact can be seen from the circumstance that the next step in this direction was made only 20 years later, namely, when D. P. Zhelobenko and M. Duflo obtained the explicit classification of all irreducible representations by indicating which subfactor of the elementary representations are equivalent to each other.

In 1956 Berezin, following I. M. Gelfand's advice, began a deep study of quantum field theory and this was the starting point of his work in mathematical physics.

In the first period of this work, from the second half of the 1950s to the mid-'60s, Berezin thought a lot about spectral theory, in particular, about scattering theory for the Schrödinger operator. There are only a few conclusive results of his in this direction, in several papers where various particular cases are considered (see [2-5]), but the observations, considerations and ideas that arose from these studies had a significant influence on other mathematicians and physicists who were on contact with him and, in the long run, led to the understanding of the spectral and scattering picture for the quantum problem for N particles that we have today (see [5a]).

At the same time as Berezin, several other young mathematicians in Russia started studying related problems (L. D. Faddeev, V. P. Maslov, R. A. Minlos, G. M. Zhislin), thus initiating the movement of mathematicians towards mathematical physics that we mentioned above. Members of this circle often talked together and rightly regarded Alik Berezin as their leader. The cooperation between Berezin and L. D. Faddeev was especially fruitful: Alik's influence on the latter was apparently very strong. Later, in the mid-1950s, the research interests of these people diverged, and their spirit of comradeship waned somewhat, but memories of that period live on in some of us.

At the very beginning of the 1960s, Berezin wrote his paper on the second quantization formalism, later presented in his monograph *The Method of Second Quantization* [6]. This formalism, long used by physicists, is based on the representation of linear operators acting in the so-called Fock space in the form of functions (usually polynomials) in certain special generators of the algebra of all such operators, the so-called *creation* and *annihilation* operators. Berezin gave this calculus a very elegant from by assigning to each such polynomial a polynomial functional on the algebra (in the case of a symmetric Fock space), so that for operations with operators (multiplication, dualization, transformations arising from canonical changes of variables, etc.) the corresponding functionals undergo transformations that are very commonplace for mathematicians: derivation, multiplication, change of variables, path integration. This method was applied by Berezin and his pupils to the study of some one-dimensional models of quantum field theory: the Thirring model (both for the massless case and the case of positive mass), the nonlinear second-quantized Schrödinger equation (see [7-9]). It should be noted that these papers had a significant influence on the development of contemporary constructive field theory. The paper on second quantization constituted the main contents of Berezin's doctoral dissertation, which he successfully defended at the Mechanics and Mathematics Department in 1965.

Berezin's study of second quantization had several important scientific consequences.

First of all, it stimulated renewed interest in the old problem of representing the so-called commutation (and anti-commutation) relations (in this connection, see V. Ya. Golodets' survey in Uspekhi [10]).

Another topic that partially arose form the study of second quantization and was developed by Berezin for many years is the general understanding of the quantization procedure. Although Berezin studied these questions from the mid-1960s, his perception is best expressed in a number of articles that appeared in 1973-76 (see [11-13]). According to the main idea of these papers, quantization has the following precise mathematical meaning: the algebra of quantum observables is a deformation of the algebra of classical observables, and the deformation parameter is Planck's constant, while the direction of deformation (the first derivative with respect to the parameter at zero) is the Poisson bracket. In the case of a flat phase space this point of view is equivalent to the ordinary one. In other cases it leads to a new meaningful theory. In particular, in his articles in Izvestia [11, 12] Berezin considered the case when the phase portrait is a homogeneous symmetric domain in complex space. He discovered a new interesting effect: the set of possible values of Planck's constant is discrete and bounded from above.

Even earlier, in the second half of the 1960s, in connection with his work on second quantization, Berezin published the paper [14] in which he studied the representation of operators in Hilbert space by using various systems of generators in the algebra of such operators (*pq*-symbols, *qp*-symbols, Weyl symbols, and the Wick symbol ordinarily used in second quantization). Note that in many aspects this paper is close to the theory of pseudo-differential operators that arose at the time and now plays an important role in mathematical physics. Thus in the work of Berezin many crucial ideas of this theory appeared independently, although, unfortunately, the significance of Berezin's work in this direction was not understood at the time.

An example of Berezin's concrete activity in this field, which led to his discovery of beautiful and important mathematical objects, is his approach to the study of the Feynman inequality

$$\operatorname{Tr} e^{-t\hat{H}} < (2\pi)^{-n} \int_{R^n \times R^n} e^{-t\,H(p,q)}\, dp\, dq\,, \tag{1}$$

where $H(p,q) = p^2 + V(q)$, $V(q)$ begin the potential, while

$$\hat{H} = -\Delta + V(q)$$

is the corresponding quantum Hamiltonian, i.e., the Schrödinger operator acting in $L_2(R^n)$. Berezin wanted to understand for which more general Hamiltonians $\hat{H}$ this inequality remains valid. It became clear that the answer depended upon the chosen quantization, i.e., on the correspondence between H and $\hat{H}$. It finally turned out that for any operator $\hat{H}$ the following inequalities hold:

$$(2\pi)^{-n} \int_{R^n \times R^n} e^{-t\,H_W(p,q)}\, dp\, dq \leq \operatorname{Tr} e^{-t\hat{H}}$$
$$< (2\pi)^{-n} \int_{R^n \times R^n} e^{-t\,H_{aW}(p,q)}\, dp\, dq\,, \qquad (2)$$

where $H_W(p,q)$ is the Wick symbol of the operator $\hat{H}$, while $H_{aW}(p,q)$ is the so-called anti-Wick symbol of this operator, first introduced by Berezin in connection with inequality (2). In the paper [15], it was proved that the exponent e^{-t} in (2) can be replaced by any downward convex function. Later inequality (2) and its generalization just described were carried over to the case when, instead of $H_W(p,q)$ and $H_{aW}(p,q)$, one considers the covariant symbols introduced by Berezin in [16], which are defined by using an overcomplete system of vectors in Hilbert space. The abstract scheme for the introduction of these symbols in [16] was later used by Berezin for the construction of quantization on Kähler manifolds. Moreover, already in the papers [15, 16], Berezin used inequalities similar to (2) in order to obtain various spectral asymptotics for the operator $\hat{H}$ for sufficiently large values for the spectral parameter, as well as semiclassical asymptotics. In particular, the paper [15] contains the first rigorous proof of the semiclassical asymptotics of the distribution function for the eigenvalues of sufficiently general Hamiltonians.

These are the main topics of Alik Berezin's research in the 1950s and '60s. We shall return to our survey of his further achievements below. But in order to assess the role of the research in mathematical physics already described, we must also discuss Berezin's pedagogical activities, understood in the wide sense. He would patiently try to develop in the physicists with whom he was in contact a taste and a feel for mathematical thinking, for the elegance of abstract deductions, and would show how to apply them to specific problems. Berezin had perfectly mastered the language and the rather loose ("galloping", so to speak) physical style of thinking, easily conversing with physicists in their own manner, thus giving a good lesson to his colleagues and pupils. He lectured in mathematics to physicists with great pleasure. A great deal more of patience and work was required to interest mathematicians in physics, to overcome their deeply rooted attitude to physics as a science beyond the limits of the understandable. For more than 20 years Berezin directed a seminar in mathematical physics and functional analysis at the Mathematics and Mechanics Department of Moscow University, sometimes by himself, sometimes together with someone else. This seminar was well known among the younger physicists and mathematicians: several first-rate scientists grew up in it, and it was the place where many outstanding papers were written. At different times he also conducted seminars in representation theory and functional analysis, and lectured in quantum mechanics, statistical physics, quantum field theory, and path integrals. His courses in statistical physics and in quantum mechanics were published in rotaprint form. Just before his death, he had started to revise the latter; this job was completed by M. A. Shubin on the basis of the notes that Berezin had prepared (see book [17]).

Berezin was very fond of discussing things with his pupils, colleagues, and friends, and he had a lot of joint papers: his coauthors in different years were I. M. Gelfand, V. L. Golo, R. I. Karpelevich, G. I. Kats, D. A. Leites, M. S. Marinov, A. M. Perelomov, G. P. Pokhil, V. S. Retakh, Ya. G. Sinai, L. D. Faddeev, I. I. Shapiro-Pyatetski, M. A. Shubin, and V. M. Finkelberg.

4. The Last Period

The Flourishing of Mathematical Physics,
Everyone Drifts to his Own Corner,
Supermathematics, Some Other Topics

During the 1960s the scope of ideas that interested mathematical physicists was constantly widening, and by the early 1970s became too wide to be grasped by one person. This period (from the mid-1960s to the mid-'70s) was truly the heroic period in the history of mathematical physics, not only in Russia but world-wide: advances in the theory of phase transitions, in general theory of Gibbs fields, the so-called "Markovian revolution" in constructive quantum field theory, new methods in the study of one-dimensional integrable systems, the renormalization group and the Wilson program for the study of critical phenomena, the birth of supermathematics (which will be discussed below), are only some of the most striking topics of the time. Of course, such a drastic expansion of mathematical physics and the increase of its proponents (which could have hardly seemed possible in the 1950s) led to a natural differentiation of their research interests; mathematical physicists progressively split up into several weakly related groups, each united around its own *maestro*. Berezin became one of these leaders and during this period worked in the much narrower circle of his nearest collaborators and pupils. This relative isolation was caused, besides external reasons, by a deep inner motivation: by that time the general approach for the construction of supermathematics became clear to Berezin, and its implementation occupied him for most of the 1970s. This approach involved a sort of "slight madness," a psychological barrier that was most difficult to overcome. This explains the small circle of people to whom Berezin was willing to disclose his plans.

We have reached the point in our exposition when we must describe in more detail the last and most significant period of Berezin's scientific carrier. The field also had other sources, but Berezin came to supermathematics, as in many other cases, from his work in second quantization. The formal calculus in the Grassmann algebra, which was developed by Berezin in connection with the second quantization formalism in antisymmetric Fock space, led him to the thought that "there exits a nontrivial analog of analysis in which the role of functions is played by elements of the Grassmann algebra" [6, 18-20], i.e., a calculus in which anticommuting variables play their role together with commuting variables. He unceasingly advertised this idea and carefully collected examples and constructions to support it.

The first construction, i.e., the Berezin integral in anticommuting variables, still remains the most impressive in the new theory, the most complicated and most difficult to really understand, although its formulation is quite simple (see [6]). This construction is closely connected to another one, also discovered by Berezin and now bearing his name, the Berezinian. In [19] Berezin developed the key case (when all the variables are odd), and in 1971 in a letter to G. I. Kats wrote out a hypothetical general formula for Berezinian, later established by his graduate student V. F. Pakhomov.

The end result of the cooperation of Berezin and Kats was their joint paper [20]. Its results are close to those of Milnor, Moore, and Quillen in the 1960s, however Berezin and Kats treat the Hopf algebras as formal Lie supergroups and indicate the relationships between formal Lie supergroups and Lie superalgebras, generalizing the exponential map and Lie theory. This paper first set the problem of constructing Lie superalgebras globally, and not only as formal objects. Two years later this problem was solved.

Finally, the last crucial new object of the theory, the notion of supermanifold, was defined by D. A. Leites [21] on the basis of an idea proposed by Berezin [22]. The construction of supermanifolds is realized along the lines of algebraic geometry (by studying the manifold by means of the local algebra of smooth functions on it) with the only difference being that in the case of supermanifolds one must use superalgebras (see Berezin's survey [23]).

In the mid-1970s, Berezin's pioneering ideas began to spread, and supersymmetry groups, i.e., Lie supergroups of transformations of "superspace-time," began to appear in the work of physicists. Thanks to the work of Yu. A. Golfand and E. P. Likhtman, D. V. Volkov and V. A. Akulov, G. Wess and B. Zumino, V. I. Ogievetsky, and many others, it became clear that supermanifolds provide an adequate language for the formulation of unified field theory. This is related to the following fundamental assumption about the structure of space-time: space-time is a supermanifold each point of which is an ordinary space-time, while the transformation group is the supergroup extending the Poincaré group via odd generators.

In the last year of his life Berezin began writing a book on supermathematics, which he was not destined to finish. The book was completed by V. P. Palamodov, using the notes and raw copies left by Berezin (see [24]).

As we approach the end of our survey of Berezin's mathematical achievements, we would like to touch upon two other topics in mathematical physics that Berezin addressed from time to time.

One of his hopes (as was the case for many others) was to construct a noncontradictory quantum field theory. Without exaggeration, it can be said that almost all of his work (on the N particle problem, quantization, superanalysis) he regarded as stepping stones to this difficult problem. He had some ideas and considerations, for example he long believed that the renormalization procedure in quantum field theory can be correctly understood in the framework of the theory of extensions: the original Hamiltonian is well defined only as a symmetric operator on an appropriate subset of Fock space, while the true Hamiltonian can be obtained as its self-adjoint extension. This idea was nicely illustrated in his joint paper with L. D. Faddeev on δ-like interaction of two quantum particles [4]. The same idea was the basis of his own paper on the so-called Lie model [25]. Here Berezin made use of Heisenberg's idea that this model should be studied in a space with indefinite metric and constructed the Hamiltonian of the Lie model as the extension of the symmetric operator to a space with indefinite metric. Many people believe that this approach may turn out to be useful in contemporary quantum theories of gauge fields, which necessarily require the introduction of an indefinite metric.

In the 1960s Berezin addressed statistical physics fairly often. In 1965 his own joint paper with Ya. G. Sinai on the existence of a phase transition in ferro-magnetic lattice structures with finite interaction was published [26]. In subsequent years Berezin repeatedly attempted to find explicit solutions for the three-dimensional Ising model, again using the techniques of second quantization (of which he was very fond and apparently regarded as a universal approach) for the purpose. Some results obtained in this direction were published in [27, 28]. Unfortunately, the significant achievements in statistical physics at the end of the 1960s and the early '70s and the related advances in constructive quantum field theory remained practically unnoticed by him. In the '70s he never returned to this topic.

Such was, in its main traits, the scientific path of F. A. Berezin.

5. F. A. Berezin's Social and Political Status and Position at the Department

General Traits of the Scientific Life at the Mechanics and Mathematics Department, Berezin as an Alien Body for the Powers that Be, the Party Rule in the Department, The Opera Story, the Letter to the Rector R. V. Khokhlov

It is difficult to assess the scientific career of Berezin, as well as that of any other important and honest scientist working in Russia at the time, outside the scientific, social, and political context within which they worked, and without taking into consideration their own social and political position.

Above we had the occasion to mention the remarkable scientific atmosphere that prevailed in the Mechanics and Mathematics Department of Moscow University at the end of the 1950s and in the 1960s. This atmosphere, despite the subsequent "tightening of the screws" (see below) did not entirely disappear until the beginning of the 1990s. At the Mechanics and Mathematics Department every year several dozen scientific seminars on various topics in mathematics and mechanics regularly functioned and about as many optional lecture courses were held. The goals and the levels of these seminars and courses could be very different, but most of them were aimed at giving additional material to the undergraduate and graduate students. To clarify the situation it is useful to understand how the traditional educational system works at the Mechanics and Mathematics Department: there is a curriculum, consisting of ten or twelve compulsory courses of lectures that usually last two (or even three) semesters. As a rule, these courses are supplemented by exercise classes, where smaller groups of students solve problems that illustrate the subject of the lecture courses. On the other hand, the seminars and brief courses mentioned above are entirely optional and are chosen by the students themselves or in accord with the suggestions of their scientific advisors (only the minimal total number of such courses and seminars is fixed by the curriculum).

Beside these educational seminars, the Department traditionally had several research seminars of the highest level, which would bring together mature mathematicians and where the latest achievements in the given field were discussed. Of course these research seminars were regularly attended by many graduate students and the most advanced undergraduates, and were an excellent school for them. We have already mentioned two such seminars: the famous I. M. Gelfand seminar and the one in mathematical physics and functional analysis directed by Berezin at the end of the 1950s and early 1960s jointly with R. A. Minlos. Another well-known seminar in mathematical physics, working at the Mechanics and Mathematics Department since 1962 and still in existence today, headed by R. L. Dobrushin, V. A. Malyshev, R. A. Minlos, and Ya. G. Sinai, was traditionally devoted to statistical physics. Mathematical physics (its geometrical aspects and the theory of integrable systems) was also the topic of the well-known seminar headed by S. P. Novikov that has been functioning at the Department for many years. All these seminars, like several others at the Department, were known world-wide, and many scientists (from Russia or from abroad) made it a point to visit them and/or give a talk there. One recalls the unique spirit of free and serious discussions prevailing at theses seminars: each participant would try to understand the speaker completely, the talk could be interrupted at any moment by a question, a clarifying remark, or by a whole flow of improvised comments by one of the participants. There was not even a hint of subordination, any participant that had something to say on the question under discussion could come to the blackboard (sometimes even during the talk) and be heard. An attitude of respect and consideration for all participants was the rule. These spontaneous and exciting discussions, often spiced with clever jokes, were truly a feast for the intellect and were perhaps the most valuable ingredients of the seminars. They often led to new understandings of the problem under question, sometimes unexpected even for the speaker, new ideas and questions would arise and later develop into serious research work. This was very important for the younger participants — teaching them the proper attitude to creative research and scientific intercommunication — and to delighted foreign visitors, bored by the stiff etiquette at their own seminars. One Italian colleague, who lived in Moscow for a long time and regularly attended the seminar in statistical physics (and sometimes, out of pure curiosity, he would go to political meetings too, they took place at the Department from time to time), joked that these seminars reminded him of political meetings at the University of Rome, while, conversely, our political meetings reminded him of dreadfully boring scientific seminars in Rome.

Among the other important aspects of Moscow mathematical life were the sessions of the Moscow Mathematical Society, especially in the 1970s and '80s, when I. M. Gelfand became its President and succeeded in remarkably invigorating its work. Each session of the Society was a carefully prepared survey of some new and interesting mathematical topic. The survey would usually be delivered by the leading expert in the field. The sessions were very widely attended, by the youngest mathematicians as well as by the more experienced researchers.

Listing the outstanding conditions for research at the Mechanics and Mathematics Department, one must mention the rich university library, in particular its mathematical part, which until recent times was systematically supplied with all kinds of mathematical publications appearing in Russia and most of the leading journals from Europe, America and Japan.

However, despite the excellent working conditions at the department, Berezin's life at the Mechanics and Mathematics Department did not proceed very smoothly. We have already mentioned the discrimination to which he was subjected upon graduation from university. The Khrushchev "thaw" gave him the opportunity to return to the University to stay. However, in his work Berezin was often faced with various external obstructions, a sort of pre-planned injustice rooted in the system itself. The discrimination and obstructions, which increased noticeably in the 1970s, often made life miserable for him. To try to elucidate the mechanism of covert pressure applied to Berezin, I should perhaps explain the traditional distribution of power at the Mechanics and Mathematics Department that prevailed until the end of the Soviet regime. An important part of the power belonged to the so-called "party bureau," the executive group of the Department's communist party organization. The party bureau consisted mostly of creatively unproductive functionaries who had found in the party a haven and a justification of their own worthlessness. These people were usually spiteful (and, as a rule, anti-Semitic) and directed their spite at the really active researchers in the department (especially if the latter were Jews). Of course, the spitefulness of the party bureau was partially balanced by certain positive rules and traditions, as well as by the influence and administrative prerogatives of the Scientific Council (and sometimes the Dean's office), which mainly consisted of real research scientists. In the period of Khrushchev liberalism and the first years of the post-Khrushchev era, when the party bosses were still in the state of relative indecision, the influence of what may be called the "scientific party" grew noticeably. This was especially so during the deanship of N. V. Efimov, a remarkable and noble personality. At the end of the 1960s, when P. M. Ogibalov (a party functionary from way back, known in the Stalin years for his active participation in various "party purges" and "denunciations") became the Dean, the party leaders united with the Dean's office and a dismal atmosphere pervaded the Department for years to come. Specifically, the party bureau according to the existing traditions could (and did) direct the life and work of any employee of the Department by means of the following prohibitions:

- Forbid pay-raise or promotion to a better position;
- Forbid trips abroad — both by private or official scientific invitations;
- Forbid graduate studies to students of researchers who displease the party;
- Forbid lecturing in a compulsory course;
- Forbid one's reelection to a faculty position for the next five-year period.

The last veto was only applied in exceptional cases (one of these led to the untimely death of G. E. Shilov).

All these prohibitions (except the last one) were applied to Berezin consistently at the Department. Perhaps at this point I should recall an amusing and typical episode in which I took part together with Berezin. One of the standard pretexts for various prohibitions was that the employee concerned did not have a "social workload" or did not perform it adequately. Here "social workload" meant an unimportant and necessarily nonrenumerated activity, usually quite dull and/or meaningless; for example, the organization of so-called "political informations" at which the person responsible would retell to students (or to his own colleagues) the contents of the latest Soviet newspaper, or the so-called "civil defense sessions," where year after year one would be told what to do if an H-bomb drops on your head, or the like. It was imperative that all employees have a specific "social workload" of this type. Of course, no normal human being could take such caricature activities seriously and people usually only pretended to be involved, or even quietly avoided doing it altogether. The party bureau was usually aware of this and looked at this deceit through its fingers, but at any moment could demand an explanation, keeping a person under stress and control, reminding him of its pervasive existence. It is typical that no serious socially useful activity, e.g., membership in an editorial board, a position in the administration of the Moscow Mathematical Society, the organization of "mathematical circles" for high school students, was acceptable as one's "social workload," unless of course this was specially allowed by the party bureau (I recall hearing the following phrase several times, "what kind of a social workload is that if he enjoys doing it?").

To come back to my story, in the early 1960s after Berezin and I had worked at the Chair of the Theory of Functions and Functional Analysis for five years in the position of junior researchers, the question of our appointments to the positions of senior researchers arose. The party bureau did not agree to this, on the pretext of the absence of any "social workload." Negotiations on this topic with the party bureau were conducted by G. E. Shilov, who, as an active music lover, was in constant contact with the opera studio functioning at the university. He decided to help us by using the studio, which was about to stage an opera whose original libretto was in the Belorussian language; he proposed that Berezin and I perform the translation. We worked hard at this for several weeks, completing a rather good Russian version (alas, this was to be our only joint work [29]) and then spent a long time with G. E. Shilov to make the text fit the music. The opera studio was satisfied with the result, the opera ran with success (our names were on the posters), but we did not get the expected appointments, because the party bureau refused to regard all this activity as an acceptable social workload. We were both appointed senior researchers only two years later, when part of the members of the party bureau were replaced. This was Berezin's last promotion till the end of his life.

There were serious problems with some of Berezin pupils as well, who were not allowed to do graduate studies. This was the case with D. A. Leites, his favorite pupil, who was a key figure in the construction of supermathematics. Concerning Berezin's trips abroad, they ceased entirely after 1975, despite an endless stream of invitations from Europe and America (one of the drawers of his desk was filled to overflowing by these invitations, as we discovered after his death). The trips that he was forbidden to go on were important to him not only professionally, but also psychologically: during those years the recognition that he so badly needed was becoming a reality.

In the mid-1970s, Berezin wrote a letter to the new Rector of Moscow University, the physicist R. V. Khokhlov, in which he described the general situation then prevailing at the Mechanics and Mathematics Department: discrimination against Jews at the entrance examinations to the University and Graduate School; the related exclusion of many honest teachers actively working in research from all important affairs of the Department, such as entrance and final examinations, and lectures in the obligatory courses; the almost total prohibition of trips abroad for an overwhelming majority of teachers; the specially organized unmotivated "failures" at thesis defenses for *candidate* and doctoral degrees for ethnic Jews; and many other aspects. It is known that R. V. Khokhlov had the intention of taking decisive measures to make life at the Mechanics and Mathematics Department healthier (his sudden death as a result of a mountain climbing accident put an end to that), and apparently Berezin's letter played a significant role in Khokhlov's unrealized plans.

After R. V. Khokhlov's death, the contents of the letter reached the party bosses, only increasing their hostility toward Berezin (the first act of reprisal was the sudden and anonymous cancellation of his already approved trip to Czechoslovakia on a private invitation).

Despite all this harassment and all the humiliations, Berezin always retained his freedom-loving and independent personality, observing the vileness that surrounded him with disgust and sorrow. Being a pessimist by nature, in the last years of his life he became increasingly gloomy, unable to see any ray of light in our dismal life of those years.

In the summer of 1980 F. A. Berezin drowned during a trip in Kolyma. His body was found and brought back to Moscow. An urn with his cinders reposes in his grave at the Vostryakov cemetery in Moscow. It is a pity that he did not live to see the present days, nor to experience the world-wide recognition of his scientific work. He would have rejoiced in the one and in the other.

February 1994

References

1. F. A. Berezin, *Laplace Operators on Semi-Simple Lie Groups*, Trudy Moskov. Mat. Obshch., **6** (1957), 371–463; **12** (1963), 453–466.
2. F. A. Berezin, *Asymptotics of Eigenfunctions of the Multiparticle Schrödinger Equation*, Dokl. Akad. Nauk SSSR, **163** (1965) No. 4, 795–798.
3. F. A. Berezin, *Trace Formula for the Multipartcile Schrödinger Equation*, Dokl. Akad. Nauk SSSR, **157** (1964) No. 5, 1069–1072.
4. F. A. Berezin, *Remark on the Schrödinger Equation with Singular Potential*, Dokl. Akad. Nauk SSSR, **137** (1961) No. 5, 1011–1014.
5. F. A. Berezin, G. N. Pokhil, and V. M. Finkelberg, *The Schrödinger Equation for Systems of One-Dimensional Particles with Point-Like Interaction*, Vestnik Moskov. Univ. **1** (1964), 21–28.

5a. F. A. Berezin, R. A. Minlos, and L. D. Faddeev, *Some Mathematical Questions in the Quantum Mechanics of Systems with a Large Number of Degrees of Freedom*, Proc. 4-th Soviet Math. Congress, **2** (1964), 532–541.

6. F. A. Berezin, *The Method of Second Quantization*, in Russian, (Nauka, Moscow, 1965); Engl. Translation: Academic Press, New York, 1966.
7. F. A. Berezin, *On the Thirring Model*, Zh. Eksp. Theoret. Fiz., **40** (1961), No. 3, 885–894.
8. F. A. Berezin and V. N. Sushko, *Relativistic Two-Dimensional Model of a Self-Interacting Fermion Field with Non-Zero Mass in the State of Rest*, Zh. Eksp. Theoret. Fiz., **48** (1965), No. 5, 1293–1306.
9. F. A. Berezin, *On a Model of Quantum Field Theory*, Mat. Sb. **76** (1968), No. 1, 3–25; Engl. Translation: in Math. USSR-Sb.
10. V. Ya. Golodets, *Description of the Representations of Anti-Commuting Relations*, Uspekhi Mat. Nauk, **24** (1969), No. 4, 3–64; Engl. Translation: in Russian Math. Surveys.
11. F. A. Berezin, *Quantization*, Izv. Akad. Nauk SSSR Ser. Mat. **38** (1974) No. 5, 1116–1175; Engl. Translation: in Math. USSR – Izv.
12. F. A. Berezin, *General Concept of Quantization*, Comm. Math. Phys. **40** (1975), 153–174.
13. F. A. Berezin, *Quantization on Complex Symmetric Spaces*, Izv. Akad. Nauk SSSR Ser. Mat. **39** (1975) No. 2, 363–403; Engl. Translation: in Math. USSR – Izv. **9** (1975), 341–379.
14. F. A. Berezin, *On Representation of Operators by Means of Functionals*, Trudy Moskov. Mat. Obshch. **17** (1967), 117–196.
15. F. A. Berezin, *Wick and anti-Wick Symbols of Operators*, Mat. Sb. **86** (1971), No. 4, 578–610; English Translation: in Math. USSR-Sb.
16. F. A. Berezin, *Covariant and Contravariant Symbols of Operators*, Izv. Akad. Nauk SSSR Ser. Mat. **36** (1972) No. 5, 1134–1167; Engl. Translation: in Math. USSR – Izv.
17. F. A. Berezin and M. A. Shubin, *The Schrödinger Equation*, (Moscow State University Press, Moscow 1983).
18. F. A. Berezin, *On Canonical Transformations in Representations of Second Quantization*, Dokl. Akad. Nauk SSSR, **150** (1963) No. 5, 959–962.

19. F. A. Berezin, *Automorphisms of the Grassmann Algebra*, Mat. Zametki, **1** (1967) No. 3, 269–276.
20. F. A. Berezin and G. I. Kats, *Lie Groups with Commuting and Anti-Commuting Parameters*, Mat. Sb. **82** (1970), No. 3, 343–359; English Translation: in Math. USSR-Sb.
21. D. A. Leites, *Spectra of Graded Commutative Rings*, Uspekhi Mat. Nauk, **29** (1974), No. 2, 209–210; Engl. Translation: in Russian Math. Surveys.
22. F. A. Berezin and D. A. Leites, *Supermanifolds*, Dokl. Akad. Nauk SSSR, **224** (1975) No. 3, 505–508.
23. F. A. Berezin, *Mathematical Foundations of Supersymmetric Field Theories*, Yad. Fiz. **29** (1979) No. 6, 1670–1687; Engl. Translation: in Sov. J. Nucl. Phys.
24. F. A. Berezin, *Introduction to the Algebra and Analysis of Anti-Commuting Variables*, (Moscow State University Press, Moscow 1983).
25. F. A. Berezin, *On the Lie Model*, Mat. Sb. **60** (1963), No. 4, 425–446; English Translation: in Math. USSR-Sb.
26. F. A. Berezin and Ya. G. Sinai, *Existence of Phase Transition of a Lattice Gas with Attracting Particles*, Trudy Moskov. Mat. Obshch. **17** (1967), 197–212.
27. F. A. Berezin, *The Plane Ising Model*, Uspekhi Mat. Nauk, **24** (1969), No. 3, 2–22; Engl. Translation: in Russian Math. Surveys. 24 (3), 1969, 1-22.
28. F. A. Berezin, *The Number of Closed Non-Selfintersecting Contours on a Plane Lattice*, Mat. Sb. **85** (1971), No. 1, 49–64; English Translation: in Math. USSR-Sb.
29. F. A. Berezin and R.A. Minlos, *The Thorny Rose*, Opera Libretto (translated from the Byelorussian), Moscow Univ. Opera Studio, 1962.

PART TWO

NATALIE BEREZIN

ELENA KARPEL

VICTOR MASLOV

MISHA SHUBIN

ANATOLY VERSHIK

NIKITA VVEDENSKAYA

DIMITRY GITMAN

VICTOR PALAMODOV

DIMITRY LEITES

On a envie de parler de son absence, de combien il nous a manqué tout au long de ces années qu'on a passé sans lui, on a envie de lui parler à lui, de se confier à lui comme on ne peut le faire avec nul autre. On a envie de lui dire combien il nous manque, qu'on lui en veut d'être parti si vite. Et puis on a peur d'être mal compris, peur d'interprétations divers et qui pourraient nuire à l image de la personne ...

We would like to talk about his absence, about how much we miss him in all these years we spent without him, we would like to speak to him, to confide in him as we can with no one else. We would like to tell him how much we miss him, that we are upset that he left us so soon. But we are afraid to be misunderstood or that one could infer from our words something that could harm his image ....

Natalie Berezin
2005

THE LAST JOURNEY*

MOSCOW – KRASNOYARSK – MAGADAN – SEIMCHAN – MAGADAN – MOSCOW

REMEMBERING F. A. BEREZIN

ELENA KARPEL

17 rue Nicolas Gargot
17440 Aytré, France

I do not know if it is right to agree to this recollection: memoirs of widows always seem to me a deficient genre, provoking the readers to seek a subtext, concealing the evidence of accounts, etc. In addition to this, the discussion here involves a person who was deeply private, whose soul was somewhat unusually sensitive, vulnerable, without skin, and therefore protected itself with secretiveness. I would have to compare it with an oyster – a delicate body hiding behind thick, externally rough, tightly closing shutters. If the shutters are open, this means that the oyster is dead. And here stands the moral question – have I the right to reveal what was trusted to me? But people want to know what kind of a person hid behind this great scientist. It is likely that History has a right to this. And here, so as not to desecrate the memory, half-truths are inadequate, because half truth for him was more unbearable than open lies.

One can hope that twenty five years is enough time for me to reflect and make my recollections just a little bit distant, elevated to that space where they belong not only to me. One thing that has not changed over the course of a quarter century: all memories return to that wicked day of July 14, 1980.

---

* Translated from Russian by Roman K. Kovalev, The College of New Jersey, Department of History, Ewing, NJ 08628, USA; e-mail: kovalev@tcnj.edu.

Beginning of the End

This was my last working day before vacation. As usual, my colleagues (I worked at the A. V. Vishnevskii Institute of Surgery), organized a sendoff for me. The nurse Olga brought her crowned cake with whipped cream; we sat around, chatted, and had a spot of tea. In a wonderful mood, I left home to pack my bags.

I recall my thoughts: "How well everything is planned and organized this year." We had plans to spend that summer in the Ukraine where we had already rented a house in a hamlet and sent the deposit. Alik's vacation was a month longer than mine as he worked at Moscow State University (Alik – that's how Felix Alexandrovich Berezin was called by his family and friends). Therefore, we succeeded in arranging for our parents to go with Natasha for the month of June to the Riga seashore, so that Alik could go to the Far East for a month where he could walk through the taiga as he had dreamed of doing for a long time. And, finally, the next day, he would fly in, and we would go to Riga to get our daughter and travel south with her. But a note was stuck in the doors of the apartment: "You have a telegram – come to the post office immediately." It was too late to go and I simply felt lazy. I called the office and asked the clerk to read me the telegram.

"I just don't know how to read it to you," said the clerk. "Maybe it would be better to come?"

"What can it be? Read it, and that'll be it. This is a message from my husband who is returning from vacation. It is surprising that he is so responsible this time."

"But no, dear, he is not returning. Rather, you'll have to fly to him..."

"But what is it?"

And she read: "We regret to inform you ... tragically died..."

It was all over. My life had shattered ...

In the first moment of shock, one feels no pain. There forms some sort of strange emptiness, and in this emptiness, I had a mission. I called my friend Galya Poliak.

"I need money. A lot."

"You got me at the right time," laughed Galya: "Borya and I

have exactly three rubles until the next paycheck."

"Take it from the Old Man" – as Borya's father, who lived with them, came to be called.

"Lyalya, have you fallen off the stove? You know that I never take a kopek from the parents" – and immediately without a pause – "Did something happen?"

"Yes."

"What?"

It was impossible to say it. Until the words were spoken, it seemed like nothing had happened. To say it was as if to betray. I was silent.

"Related to Alik?" asked Galya curtly.

"Yes."

"Is he alive?" Galya knew what she was asking. It would not have been the first time that friends had died in these damned expeditions.

"No."

"Do not move. I will come immediately. I will bring everything. And you drink some vodka."

"I don't have any vodka."

"I will bring some. In the meantime, make yourself some tea with lemon."

"I do not want any tea."

"Can you make some good strong tea for me? And gather things for the road." Galya was great. The scariest thing is to be without purpose or action. When something falls on you, taking away your breath, it seems that you will suffocate in pain momentarily. And any activity distracts you, leads you to the side. Galya did not come alone but with my friend, Lyonya Nevler, along with his wife. I asked why. They said something like, "When a person is in pain, they have to be surrounded by people, and the more the better." What was the difference? Around me was an impenetrable emptiness.

Galya poured a pile of money on the table. And I mean a mound. Because the Old Man kept money in a savings bank, and I planned to fly at dawn on the first flight, she ran around to all the neighbors in her tenement and borrowed money against the Old Man's savings, thereby collecting all of the cash necessary. Aside from the large bills, there were crumpled up 10s, 5s 3s, but I do not recall any singles. Filling me up with vodka and tea, I was laid to sleep. Lyonya remained on guard. In the morning, joined by Galya, we went to Domodedovo.

In the airport we were attended to with sincere understanding, taken through everything without having to stand in line, and told that they would send me off on the first flight, placing me in the service sector since there were no free seats. And only in the last moment when I paid the money for the ticket, did the cashier suddenly say to me, "Give me the telegram."

"Which telegram?"

"Well, the one from Magadan, which reports your husband's death."

"But I do not have that telegram, since I did not go to the post office."

"Without a verification document, we do not sell tickets. Magadan is a restricted zone. You not only need the telegram, but it also has to be officially certified."

"Certified how? Why?"

"Because it is easy to forge a telegram... Who knows who would want to go there..." "People do not usually go to these places willingly," Lyonya tried to intervene, "Usually people are sent to these places against their will."

"Comrade, I feel for your grief, but remember that I am acting in your interests. Even if I violated the regulations and sold you a ticket, in Magadan your friend would not be released from the airport but in the best case scenario, would immediately be sent back to Moscow. Otherwise, they would hold her in a detention cell while her documents are verified. And I am not even saying that for a professional mistake like that I could receive a reprimand from the Party bureau."

"The law is the law," so we went back to the city.

Today Domodedovo is a first-class international airport served by a luxurious electric train, which will take you almost from the runway of the airport to the newly restored, elegant Paveletskii train station in forty minutes. But back then ... then one had to drag oneself to the train station, and then travel in a rickety commuter train an hour and a half to the city. That part of my being which retained some ability to understand felt sorry for Galya and Lyonya, who had to fiddle with me for so long, but I myself was unable to feel these stupid, lengthy journeys. Time seemed to stop ... no, it **has** *stopped. "Life is eternity, eternity is an instant" – in certain situations this is not a poetic metaphor, but pure reality.*

Everything was like a dream: for some reason, we went not to my local post office but elsewhere, where there was some sort of an office where, supposedly, they registered all telegrams which came to Moscow. I was seated in a spacious hall, multitudes of people were passing back and forth, Galya and Lyonya were running around somewhere arranging something. Finally, we received the necessary paperwork and we went back.

I flew out on the evening flight.

Moscow – Krasnoyarsk

I do not know how it is now, but then a flight to Magadan took ten hours with one midpoint layover. The plane was packed to the brim. Because there were not enough seats, some young people, happy, youthful tourists flying to Krasnoyarsk, tossed down their backpacks near the entrance and situated themselves on top of them and, the entire way, sang and played the guitar.

> *" I am chasing, I'm chasing mists from the Arctic*
> *And the táiga's fragrance rain-imbued"*

sang the youths in the aisle. [1]

And I was traveling for a body – the body of Alik.

* * *

How did we meet? This was nine years before that day, on New Year's Eve. It is just unbelievable that this occurred nine years ago... A posse of mathematicians, based on a long-established tradition, left Moscow for the winter break at that time, to the hotel outside

[1] The refrain from a popular song of Alexandra Pakhmutova, reputedly Brezhnev's favorite composer. Her Komsomol songs were wide-spread in the 1970s and 80s. Translated by Prof. A. Liberman. – Editor's note

of Kalinin. Galya invited me to go with them. I felt shy: they were all smart and had known each other for a hundred years; therefore, I thought I would feel greatly out of place. But my girlfriend did not want to give in and, so as to entice me somehow, began to relate to me in great detail about everyone who was coming. Her description of Alik, I recall, included: "Don't even look at this frozen professor, since you won't thaw him." But how could I even think of flirting as I was more concerned about not embarrassing myself!

When we arrived at the hotel, those friends who had arrived earlier all poured out to greet us. A loud group of ten people all congregated in the lobby. At the side there appeared a man with a beautiful grey head of hair and a pale thin face. At first I didn't even understand that he also belonged to our company. It appeared to me that we blocked his path and did not permit him to pass. Much later, I came to find out that this was a very characteristic posture for Alik – to be together with everybody and at the same time, separate. Then out of the crowd swam Galya: "Meet Alik. And this is my friend Elena, also known as Lyalya." We greeted each other, and I noted with surprise how the pupils of his blue eyes swiftly dilated. The eyes became almost black. I also thought, "What a lively reaction in this frozen professor."

The next morning, the entire company went off to go skiing. Galya perpetually made fun of me, saying, "Lyalya does not ski, Lyalya just stands on skis." Unfortunately, in this joke lay a dose of truth, coming close to being absolute. So as not to be a bother, I did not go with them and remained to walk around the hotel. There, as if out of the blue, appeared Alik who suggested we ski together.

"It will not work, as I ski slowly."

"It's OK – today I also am not disposed to break any records."

The whole situation ended with him getting cold and, so as to warm up, he began to run away from me at a quick pace some fifty meters, and then, just as briskly, run back towards me. Like this , he ran back and forth until the return of the whole group from skiing, all of whom were greatly surprised with the marvel. When all was revealed, there was no end to the jokes.

After several days, we were all returning to Moscow. Right before the breakup of the party, already in the subway, Alik suddenly said, "I will give you my telephone number. Call me sometime."

Such an inflexion (now I understand that it came from shyness) was not at all satisfactory to me.

"You know, I, of course, am for the equal rights of gender. But at the same time, I prefer that the initiative comes from the male. If you are in the mood, call me yourself."

"I do not have anything to write down your number – I am afraid I'll forget."

"Well, if you'd want, you will find a method to contact me."

And with that, we parted. Quite some time passed. Suddenly, our secretary came into the experimental surgery room, saying, "Elena Grigorevna, can I interrupt? Someone has been trying to reach you all day long."

I ran to the telephone, and Alik was on the line. "I have two tickets to the conservatory today. Do you want to go?"

I was surprised, elated, and confused. Our group at that time was developing a method for heart transplants and operations were very long. Furthermore, after their completion, and having to walk the dogs, sometimes we even stayed all night.

"I have a long experiment, and I absolutely do not know when I will be free. It's a shame, but I am afraid I might stand you up."

At that time, I did not know that Alik understood this in the best light: A person involved in her studies – that is what he valued!

"It's OK. You will not stand me up. I will leave the ticket with the receptionist. What is your last name?"

"Karpel," I said.

A strange silence in the receiver ensued. Only later I found out that my last name sounded almost like the last name of Alik's previous wife, "Karpova."This was a weird coincidence.

The dog quickly died, and I made it to the conservatory.

We started dating. It happened sort of like this. He called once a week and invited me to walk outside the city. On returning home, full of despair and physical exhaustion, I cried on the telephone to Galya: "We walked the entire day, fell into the swamp, then climbed through shrubbery, dusty spider webs filled my nostrils and blinded my eyes, and on top of everything, I had to converse on intelligent subjects! And I was such a fool, such a fool! That's it, he will never call me again!"

"Well, of course, poor Alik has no one to converse with. He invites you exclusively for philosophical debate!" laughed Galya.

Up to then, I had never encountered anything like this. If anyone was interested in me, there were, naturally, calls every day, dates at a minimum of two to three times a week. And here... Only later, when Natasha was born, I saw with what difficulty Alik forged out time for her, and understood then that, in reality, he had courted me like crazy. But at the time I had absolutely no idea what to think. But there was a feeling that I was lifted up by a roaring stream that could not be resisted. There was no question that I would even refuse or miss a date.

Such was the situation for some half a year. Then, suddenly, like an avalanche, our relationship leaped to a different level. Some of my doubts started fading away. But in no way was it possible to say that everything became lighter and easier. In any case, just when it appeared to me that our relations somehow stabilized, Alik suddenly stated:

"We have to break up ..."

"What? Why?!"

"You are only wasting your time with me. I can never marry you..."

"Excuse me, what's this about marriage?"

"...because I cannot leave my mother and no one can live with her but me."

"And did you ask me if I want to marry you? If I consider the time that we spend together a waste? And what I expect and don't expect from our relationship? In fact, as it happens, the expectation of marriage is in no way attractive to me. I want everything to remain as it is."

"Well, decide for yourself. I warned you..."

* * *

His mother was Esfir' (Esther) Abramovna Rabinovich. Even in the depth of her old age she remained beautiful, that expressively Semitic beauty which is encountered in the paintings of El Greco and Goya. She was never married – she threw out Alik's father when she was in her fourth month of pregnancy. To my question posed to Alik, "Why?" the answer was, "It is painful for me to talk about it." During the course of her entire life, E. A. refused to accept any sort of assistance from Alexander Berezin, who gave the child his name and never gave up rights to him.

Once I asked Alik, did he and his father ever know each other? He answered:

"When I was three and a half years old, he brought me a wooden truck as a present. I played with this truck for a long time. That was the last time he saw me."

Out of this, I came to the conclusion that Alik briefly saw his father later. Perhaps it was when he was already working at the university and was supposed to go on assignment to Mongolia. According to the former rules, to leave the country even for a short while, one had to present written permission from the parents or a certificate of their death. Age notwithstanding. It did not make any difference whether or not the parents, during the course of their lives, took care of the child. Once E. A., laughing, related to me:

"When I called Alexander a quarter century after I threw him out and asked him to come and sign a permission slip for Alik's departure abroad, he could not believe his ears that I would invite him to step over the threshold of my home."

When I met Alik, his father was no longer amongst the living.

* * *

E.A. was born into a family of Kishinev intelligentsia: her father had some sort of an important position in the high school; and her mother was a "midwife and smallpox vaccinator" (the door plaque with such a statement is still kept in my desk). As many were in those days, both were involved in the revolutionary movement. Kamenev, Zinoviev, and Pavel Yushkevich [2] (Pavlik, as he was called in the

[2]Prominent *Bolshevik* revolutionaries. –Editor's note

family) were friends of the household.

After the revolution, the family moved to Moscow. The parents, by that time, were no longer young – the banner of fighters for the bright future had been taken up by a younger generation. In her youth, E.A. dreamt of becoming a pianist and studied at the Vienna conservatory. (When I met her, she was embarrassed to play amongst outsiders; she still enjoyed performing when she was alone at the apartment. A brilliant piano, a Steinway, was the only luxury that she permitted herself to have in her life. This piano, even now, stands in our Moscow apartment.) In the 1920s the passionate builder of communism E.A. decided that the great cause should be assisted with something more real than classical music. Then she finished the medical institute and became a doctor. Along with practicing medicine, she became involved with scientific work in the area of histology, for which she received a prize and in which defended her dissertation.

But, then a tragedy befell her: the brother of E.A. was arrested (he was the minister of either industry or transport). When the news of his arrest broke, she grabbed her infant son and disappeared into the Russian hinterland. When they came for her mother, she answered to the question of the NKVD [3] agent:

"Where is your daughter?"

Grandmother, an old revolutionary, answered:

"You can cut me to pieces, but I will not answer."

"Leave the old woman, we will take somebody else," said the second *chekist*. And then they left.

Later on, it became known that this indeed happened. During the Khrushchev "thaw," E.A.'s third-removed cousin suddenly appeared, someone whom she had not seen since childhood, and E.A did not recognize her. She said to E.A.:

"Had you not have run off, I would not have spent all of those

[3] The Soviet acronym for the name of the secret police at that time. The Soviet secret police has changed acronyms like a chameleon. It started out as the Cheka, and then became the GPU, the OGPU, the NKVD, the NKGB, the MGB, and finally the KGB. Even today, however, people often simply refer to the secret police as "the Cheka" and the secret agents as "the Chekists." – Editor's note

years in the camp."

E.A. found a job in a god-forsaken place in Russia's heartland, at a tuberculosis hospital. There, she withdrew from clinical work and began to work as a pathologist-anatomist...

I think that these childhood years greatly impacted Alik. His reserved nature of detaching himself from society, his silence, secretive nature, and deep pessimism were, in my opinion, the result of a "happy Stalinist childhood," when he and his mother were in semi-legal status, and it was necessary to live undetectable lives whenever possible. Alik relaxed only in nature, when hiking with friends. There is one sole photograph, the only snapshot of Alik's unrestrained laughter, made when he and Valerii Nikolsky were in the northern Urals, wearing some sort of monstrous windbreakers, endlessly happy...

1940 brought a bit of light to their exile. Mother and son had just come back to Moscow when the war began.[4] They, with her elderly parents, were evacuated to Alma-Ata. There, out of the family of four people, only E.A. worked. Only based on several slips of the tongue could I imagine what they had to withstand – it was not in Alik's character to complain. I'll bring up two examples. Once, when we were living together, I was in an elevator with our neighbor. In my hands, I held a huge pumpkin.

"You eat pumpkin," thoughtfully stated the neighbor.

"Yes," I answered. "Alik and his mother very much like pumpkin with millet porridge."

I did not even know about the existence of such a dish previously; they had introduced it to me.

"It is immediately clear that they do not know what hunger is. With such pumpkins we fed ourselves during the entire war. Now I cannot even look at it," said the neighbor.

I related our dialogue to Alik, and he smiled. "The next time you ride in the elevator with him, tell him he's the one who does not know what hunger is. For us, such a pumpkin was considered a huge

---

[4]The author means WWII, which in the Soviet Union started with the German invasion on June 22, 1941. –Editor's note

feast..."

They were among the first to return from the evacuation . Once I asked, "Was it very difficult in Moscow near the end of the war?"

"It was more difficult afterwards."

"Why?"

"They stopped letting people into the subway barefoot, so you had to sneak past the *militsioners.*[5]"

During the winter he walked around at home wearing socks, but during the summer simply went barefoot. He ignored slippers as a misconception. And looking at those bare feet, I always thought, "Greetings from the hungry childhood..."

After the war, things got a bit better. Alik grew and studied well – all of his diplomas were diplomas with honors; he was always a straight-A student. While in school, he began to participate in the mathematics club at the university, which was led and advised by Dynkin. E.A. defended her doctoral dissertation. Materially, things got better. They bought a piano and began to actively assemble a family library.

There is no question that E.A. was a brilliant pathologist-anatomist but she had a difficult character. She was accustomed to her word being "the law," and was absolutely intolerant of the opinions of people around her. Once I asked Alik why E.A. refused to have clinical practice. "Because she finds it impossible to talk to the patients," he answered.

When I came to their home, E.A. was already retired. Alik remained her only light in life. There was no one else around. Sometime in her childhood there had been girlfriends, but with those who did not die, she had ended relations. The only person with whom she could share any problems was her son. Unfortunately, the birth of her granddaughter also did not beautify her life. She told me on a number of occasions: "Natasha is happiness for you, Lena; only my son exists for me."

Once I asked Alik, "Look, E.A. was a beautiful and interesting woman. In her youth she probably attracted the attention of many

[5]Policemen – Translator's note

men, didn't she?"

"Well, she did, but mother did not want anybody. One person was trying to get to her for five years, but later even he gave up."

"But if she had not been alone, you would have left the house?"

I will never forget how he looked at me. Well ... mother and son have their own secret bond. I do not want to and do not have any moral right to touch upon it. When Natasha was born and we began to live together, I did my best to build a new architecture within the family and create a pleasant communal life. Unfortunately, to no avail. E.A. could, right on the spot, while we sat peacefully in the kitchen while awaiting Alik to come to lunch, say to me, "Now, Elena, while Natasha is little, you are necessary for us so as to take care of her. But when Natasha gets older, and we have no need for you, it would be very nice for the three of us without you."

The end was inevitable: I wrangled a cooperative apartment [6] out of my institute, grabbed Natasha, and left to live separately. From then on it only got worse for Alik – now he had to rip himself into three parts: between mathematics, his mother, and us...

* * *

Once I said to Alik, "I stand in the fifth place for you."

"How did you derive that?"

"In first place is mathematics, in second place is your daughter, third and fourth is divided between your motherland and your biological mother, and I receive the "honorary" fifth."

"Well, if you like to be so."

"No, I do not like this at all, but fact is a fact."

"Fact? It depends how you look at it."

"And how should one look at it?"

"Lay this out not in planar geometry, but in three-dimensional space."

Well, on such mathematics, I did not have enough brains. I happily responded, "So this means that in some space there is no one

[6] Something like a condominium. In the Soviet Union it was very difficult to get them. –Translator's note

between us? This suits me."

"Me, as well," smiled Alik.

When Natasha was born, he returned to the issue of marriage.

"Perhaps, for the benefit of the child, it would be better if we formally marry?"

I have to note that in his voice rang an obvious note of unwillingness.

"Why?" I answered. "My parents were never registered. More precisely, they got married when I turned sixteen years old, and I needed to receive a passport. And even this was not necessary. For my own self-esteem, I do not need to be registered – I have already been married once. As for you, in addition to our common "prison of peoples," in which we all live behind the Iron Curtain, you sit in a personal cell, created by your mother's love. You yourself are afraid of hanging on yourself extra marital chains. No, let E.A. live for as long as it is inscribed in the Book of Life, and then you will live by yourself, breathing at least a bit of relative freedom. And, if later you come to me and Natasha and say that you want to register, it is then that we will evaluate the situation and see if I will agree to put a new stamp into my passport."

Alik hugged me silently. To my mind, he was grateful to me. But I know that he always had a predilection that his mother would outlive him...

As for the documents concerning Natasha, from the moment of her birth, Alik determinately put into his head that her papers had to be in an ideal state. He himself went to ZAGS [7] in order to officialize his parental rights. From there, he was sent on numerous occasions to obtain additional paperwork, and they later demanded a written statement from me that Felix Alexandrovich Berezin was recognized as the father of my daughter. But, later even this turned out to be not enough. I was required to come in person. So Alik led me with an iron hand to ZAGS. When we came the supervisor greeted him like an old acquaintance.

"Instead of having you come in so many times with various papers,

[7] Office of Civil Status Registry and Documentation. –Translator's note

it would have been easier to simply get married. Nothing prevents you from doing so – neither of you are at the moment married."

Apparently she was concerned for my interests!

"No, we had better go our own way," I replied.

"You are sure about this?" the supervisor said with skepticism.

"Yes, I am sure..."

I was never Alik's official wife, and I became his unofficial widow...

* * *

In my memoirs, the pronoun "I" occurs often – I admit this. But I do not write a scientific treatise about a renowned scientist, mathematician-theorist who had died on the forty-ninth year of his life. I attempt to convey my subjective recollection about a person, reconstructing the atmosphere in which he lived and attempting to relate what occurred in his " first circle."[8]

* * *

Alik had his own very stern demands in everyday life. Nothing in any possible way could disturb his work. For example, when Natasha was born, he said to me, "Arrange things in such a way that it will be quiet at night, otherwise I do not think clearly."

In general, there were complications with sounds. He had an acute sense of hearing. Without straining himself, Alik could hear the conversation amongst passersby who walked at a distance of 1.5 to 2 meters from him.

"The KGB is dying to have you," I used to joke. "If they knew, they would find applications for your talent."

In response, Alik would become irritated – Soviet security forces were among the rare themes for which he had no sense of humor. Before meeting him, I used to think that an acute sense of hearing is a wonderful thing. But, as it turned out, possessing such a distinction

[8]A reference to the Solzhenitsin novel *The First Circle*, the title of which is based on a quotation from Dante. The novel describes three or so days in the life of the occupants of a *gulag* camp in the Moscow suburbs, the Marfino sharashka. Many are technicians or academics. The manuscript of *The First Circle* was seized by the KGB in 1965. –Editor's note

has its own significant inconveniences. In the apartment on Vinnitskaya Street, Alik could not sleep in the room where the window was on the side of the boiler since the humming disturbed him, despite the fact that they lived on the ninth floor. Once the neighbors, who lived one floor down, hung up a chime clock on the wall against which was pressed the couch on which we slept. This ended in a tragedy – Alik lost sleep. I had to go to explain. The neighbors, pleasant people, could not understand anything, but accepted my plea and re-hung the clock on a different wall...

As for food, clothing, and comfort, his demands (if you could actually call them demands) were minimal – being nourishing, fits well, not cold – no more than that. The sole complication presented itself as head-wear. In addition to the fact that, for a male, Alik did not have pronounced facial features, the brain part of his head was significantly larger than average.

"Why was it necessary for humans to evolve so far from monkeys?" I used to lament, when it continuously appeared that his winter hat, which was procured with such difficulty, turned out to be too tight. For skiing, he kept specially knit woolen hats for decades.

The question of fashion did not exist for Alik. In the 1970s, when various types of nylons and similar jerseys had just come in, he refused to wear clothing made out of them. Now, we all know that synthetic fabrics do not breathe and often cause allergies. But at the time this was perceived as one of his idiosyncrasies. He wore cotton shirts in the summer and woolen sweaters in the winter. It made no difference whether these were old, worn-through, and discolored items from being washed with other things. For his winter hiking trips, he had a pair of woolen underwear from those which had been sent by the Allies in the Lend-Lease Act, along with canned meat, for the Red Army.[9] I have no idea how this underwear came into the family, but Alik greatly valued it, maintaining that such items one could "no longer get." I was forbidden to wash it because, due to its old age, it could fall apart or lose its warm qualities. I was

[9]Lend-Lease was the major United States program in 1941-1945 which enabled the United States to supply the Soviet Union and other Allied nations with vast amounts of war material. –Editor's note

allowed only to air it out on the balcony and then later stuff it into a sack and toss it into the loft until the next winter outing. When I tried to rebel saying that this was " just dirt" Alik answered, to my dismay, "On the second day of an outing, I become no cleaner than this underwear."

He had his own understanding of what is seemly and what is not. I do not recall having ever seen him wear a necktie. The collar of his shirt was always unbuttoned. "Otherwise I cannot breathe."

But, at the same time, he never allowed himself to go around wearing an unbuttoned blazer or coat. Once one winter we rode in the subway. The train stopped amidst two stations. Due to the absence of ventilation in the car, it became unbearably hot. I advised, "Unbutton the coat."

"No, it is unseemly. So what if it's hot? It is necessary to persevere."

And this was all of Alik – a free spirit and a buttoned-up soul...

Having moved to Vinnitskaya Street, I became acquainted with their household. In the kitchen, in the drawer for tableware, amidst the rather unpretentious steel spoons and forks, one tin spoon stood out in its unattractiveness. I asked whether it was possible to throw it away.

"Oh, yes, you would eat with silver or, in an extreme case, with cupronickel; who cares that this spoon is light and of awfully convenient size and form ... "

Indeed, the spoon, as it turned out, was surprisingly convenient. Natasha used it all of her childhood, and I carry it with me from apartment to apartment, from country to country, and now I intend to bring it to my granddaughter...

Alik did not smoke or drink. Now this was not a determined resolution; he simply did not feel a special need to do so. He did not like coffee, but while working all day long, would drink strong, freshly brewed tea with lumps of sugar, which he nibbled on, something that to me was an apparent reminder of the war years.

When I moved in with them in Vinnitskaya Street, he gathered his friends to celebrate the occasion. I ran about the shops, "procured" various delicacies, and prepared all sorts of food. The guests, as usual, all arrived late. Finally, we sat at the table, and, suddenly, it came to be known that there was not a drop of anything alcoholic in the house, and the stores were all closed. All, to varying degrees, were perplexed. I recall how angry Nikita was. She had spent the entire day skiing outside the city, anticipating how at Alik's, she would relax with pleasure with a glass of vodka. But such a thought never entered Alik's head, and I, in light of my inexperience, had not anticipated that I would also have to take care of this.

"And you did not kill him after the guests departed?!" asked Sveta, when I told her about the disaster of my first party.

"For what? He's a whole and sincere person and, therefore, cannot expect to offer his friends something in which he finds no pleasure."

Our daughter does not smoke. Nor does she drink wine, cognac, or coffee. Living in France, this is a challenge. But apparently the father's genes turn out to be stronger than the tradition of the country.

To imagine Alik in an atmosphere of luxury would have been absolutely impossible. Once one winter he agreed with his friend to work for two weeks at his empty dacha. I once got a free day, and so I decided to go and visit him. The dacha was in the village of Old Bolsheviks, situated along one endless street, down both sides of which stood absolutely identical two-story wooden houses. In those times, these dachas seemed the peak of luxury. I walked along the middle of the street. "It cannot be that Alik found himself in such chateaux."

I walked for a long time. And then I saw a dacha with a slanted wall and caving-in roof. I did not need to look at the number of the house. I knew that I had come to the correct address.

Alik had unprejudiced opinions about everything. During the same visit, after lunch (I brought homemade food, because I knew that on his winter trips, he lives on kasha, pasta, and canned food), we lay upon something like a couch, laughed, and I related the latest Moscow news, and later noted, "In general, it appears risky to come here without warning you. And what if you were not alone?"

"But I, in reality, am not here all by myself. I just don't know whether it is a she or a he."

"How do you expect me to understand that?"

"Here, there are cracks the size of a finger in the walls. Whether I heat the house or not, by morning it is terribly cold. So at night a mouse comes to visit me, sits on my chest right at that place where your head now rests, and we heat each other."

I screamed and jumped from the couch.

"You prefer that it be a woman?" laughed Alik. Well, what can you say to this? I returned to my previous spot.

* * *

As much as everyday life did not concern him, the opposite was the case when it related to intellect. His choices for his circle of friends were determined by the breadth of knowledge and the ability to think in a nonstandard way of a friend-to-be. Here there were no compromises. Once the decision was made – with whom he wanted to associate and with whom he couldn't – it was impossible to move Alik from these positions.

But at the same time he was also always interested in an interesting person, any person. While on vacations he used to like meeting new people from different social backgrounds. He didn't have a drop of snobbism. It happened, at times during the summer when we vacationed in the village, I would go along the street and see two men talking in the distance. I would come closer – one of them would be Alik. If he wanted to talk to somebody there was little that could hinder his way.

Once, my former father-in-law, with whom I maintained warm relations, was passing through Moscow. Alik declared that he wanted to meet him. I was highly surprised and asked why he needed this.

"Based on your stories, this is an outstanding scientist and person (which was absolute truth), and life does not offer many chances to converse with such an exclusive character. Hence, I don't want to lose such an opportunity."

And despite the fact that this was in the middle of summer, while we were living at a dacha near Sotrino, and we did not want to drag the one-year-old Natasha to scalding hot Moscow, because there was no one to leave her with, we went to the city. Alik was pleased, and when Grigorii Leonidovich Lempert did arrive there soon after, he was sorry that he only had one chance to meet with him.

There was no such thing as putting too much energy into trying to have an intense discussion. In general, such a form of communication for him was extremely important. But in contrast, everything that reminded him of everyday chatter, Alik could not stand.

For example, there was a house in Moscow which I liked to visit. The lady of the house was intelligent and good-spirited, but she had one weakness – she desired to create at her home something like a salon. I just don't know how Alik picked up this nuance from my brief discussions, but there were no forces that I had that could drag him there with me. The lady of the house so very much wanted to get the prize of "ingenious Berezin!" But Alik said, "She will not get me" – and all of my requests were in vain.

* * *

Maintaining his physical form was a primary part of his personal hygiene. He did not exercise when he woke up, but tried to go for a jog at least twenty minutes before going to sleep. It is only now that just about one of every two people jogs, but then he was the first amongst my acquaintances who began to run. To spend one day a week outside the city was just as important for him as it was to all of us to go to a bath house when the hot water was turned off in our homes. Under the slightest of pretenses, he would leave to wander around, go skiing, and ride a bike whenever he could – around the city and during distant excursions. He was capable of long-distance walks and lengthy skiing journeys. As young people, they went out in large groups, but in later years many slowly began to drop out. After the death of Valera Nikolsky, the more common companions of Alik were Nikita Vvedenskaya and Victor Palamodov.

To his sporting inventory, he paid far more attention than to his clothing. The front entrance was overloaded with skis of different types – downhill, short-distance, cross country, long distance, and winter outings. He greatly valued his bicycle and, thus, did not keep it in the hallway of our building, but there was no free space in the apartment; therefore, Alik tossed out the back of our sofa, on which we slept, and in the free space, he constructed a stand for the bicycle and camouflaged it with Vietnamese straw. It all turned out merry and amusing. Our guests were enticed to find the bicycle in the apartment. This was an almost unfailing trick for Alik – no one could find it. (When Alik wanted to and found time, he could be quite useful around the house – we never had to call in an electrician or repairman.) I used to "complain" that I slept hugging not Alik but the bicycle. But I used to encourage him in his maintenance of physical form, "Well, of course, inside a healthy body there, is a healthy spirit."

"And in a healthy body, there is a work-capable brain," – he replied in absolute seriousness.

Later, when the court-appointed doctors in Magadan were performing the autopsy, I was told, "Well, they say that Siberians are healthy, but I do not recall the last time I saw a man over forty years old with a heart and veins so untouched by age, as in this Muscovite."

Alik used to say to me with surprise, "You consider yourself to be a cultivated individual, but physical culture is such a vital part of a cultured society – so how can you not take good care of yourself?"

"Well, what do you expect – I grew up a sickly child in a Jewish family in which, not for medical reasons but by mother's request, I was excused from physical education in school during the course of all ten school years."

My physical ineptitude led to our inability to go out camping together. And in the realm of our everyday life, we had disagreements due to our different backgrounds. In accordance with Alik's mannerisms, there were traits that very much disturbed me. For example, he did not leave tips. Once I said that our hungry childhood was long behind us, and it was time to behave like a respectable person who can support his material position in life. He answered, "No one gives me tips for my work, so why do I have to do it for others?"

But Alik was by no means stingy. He easily lent money, if money was around, and was not angry when it was not returned.

"It is clear that he has no money," he said in one of such circumstances.

As for myself, I was raised in a well-to-do family. My childhood and teenage years were spent in Riga where, at that time, gallant behavior had not yet been erased. When we moved to Moscow, for a long time I could not get used to the rudeness, which it seemed had soaked into the air of the capital. Alik would periodically get some of this "air." Once, an entire storm came about because he did not give me his hand when we were exiting a trolleybus. At a loss, Alik tried to to justify himself, saying, "If we had been on a hike or it was necessary to overcome some sort of an obstacle or you were in the city walking on crutches, then, of course, I would have helped. But in the present case, why?"

"Simply put, you're a certified representative of Muscovite rudeness and sassiness. If you went to work overseas, then before they found out that you're a genius scientist, they would think you were an unrefined commoner," I would plead with him.

What is most amusing is that in his behavior, Alik turned out to be much more contemporary than myself. When Natasha was in France, attending boarding school in a neighboring town, and I came to bring her home for the summer holidays, a young lad, who was obviously in love with her, came to see her off. The three of us walked toward the car, and Natasha and I dragged her heavy bags behind us while the embarrassed young man walked alongside us with empty hands, being indecisive in offering his assistance. Out of fear of underscoring my age, and for Natasha to not have to touch upon the issue of equal rights of women, we chose not to solicit his assistance. We crossed the students' sports court, there was not a single cloud in the sky, and I walked and thought, "Alik, if you can see us from above, how you now must be laughing at my misunderstanding of foreign rules of good behavior!"

* * *

Yes, many things separated us. Quite likely, there were more factors that divided us than united us. We both loved our work – were absorbed by it – and this was, already at a minimum, half our lives. But if I could describe what I did in a very general way, for me it was absolutely impossible to understand what occupied his mind. I loved noisy company, merriment, and dancing. Alik believed that real socializing could only be *tête-à-tête*, and that he just had no time for any sort of sitting around making small talk. On the outside, he appeared easy going, the movement of his body was somewhat restrained, but internally he always ran as if he felt that he had little time left, that he was late... He loved concerts, theater, and the cinema but almost always said, "Go yourself and then you tell me about it." When I came home, he and Natasha were usually already fast asleep. The next day during supper, I would describe what I saw.

During the October months there were five to six birthdays to attend of the people who were closest to him – his mother, Nikita Vvedenskaya, Valera Nikolsky, mine, and several other friends. Alik adored these celebrations and took part in them with pleasure, but every time he grumbled, "October – an awful month, there is absolutely no time left to work."

And so, what united us? I think that if we had met early in youth, our union would not have lasted. But we met when we already had several life experiences behind us and a lot of sadness. We had already learned how to be patient, and not to overemphasize those things that divided us. We valued what united us. We were like two molecules which tend to bond and hold on. We wished to lean on each other, to warm each other up. Out of all human qualities, he valued kindness most of all, considering this to be a gift which one encounters more rarely than genius. Once he said, "In my life, I have only encountered ten geniuses, but kind people, only two, the second being you."

In reality, this was not true. If one is to give full disclosure, our friend Sveta had said to me once, "No one has the ability to support and comfort someone as you do when you want to, but no one tosses someone aside as coldly and heartlessly as you, either."

My relationship with him was something special, even there in the lobby of the Kalinin Hotel where we first met. In the depths of Alik's dilated pupils, there appeared to me a call for help, and I desired to offer my hand as if to someone drowning. And then during our ten years together, the feeling never left me that, in the entire wide world, only I could help him get through life... Oh, well...

When we decided to unite our fates, it was necessary to also unite our libraries. In their three-room apartment, along almost every wall, stood shelves packed with books. The books were not only read but worked-through and analyzed. I could not cease to be amazed not only at the breadth of Alik's knowledge in various branches of science, literature, and art, but also at his persistent interest in any given trivial information. Such an impression he left not only on me. For example, recently Viacheslav Vsevolodovich Ivanov was recalling the engaging conversations he had with Alik during their walks, and how once, when they accidentally met in Dubna, they discussed with mutual pleasure the meaning of the number "two" in various descriptions of the world, from mythology to contemporary science. To contemplate and analyze was just an everyday part of life for Alik, like breathing. As a result he had his own specific view about every issue, and it did not just fall out of the air but was based upon careful thought.

But I will return to the story of the library. It consisted of several parts, the largest being mathematics and physics – these books were later partially taken by Alik's students, and the rest I donated to the University. There was a section on medicine and biology; here E.A. and I had many identical copies. Alik loved and collected fairy-tales, and there was one shelf packed with fairy-tales of the world. In addition, there were four "hidden" shelves which were locked with a key. Pre-war and pre-revolution literature was kept there. Most of the books lacked covers and title pages – traces of Stalin's era when, being cautious of searches, people concealed the names of authors who had been declared "Enemies of the People."

In the section of literary fiction, we also had many duplicates. Some of them we traded and the rest we sold, but we balked at Pushkin.[10] We each had identical ten-tome sets, only differing in the color of the spines – he had brown, and I had blue. Each wanted to keep our own. As a result, we retained both. The books stood on the shelves, one under the other, and each read their own. Natasha knew the history of both editions. She asked me once when she grew

[10] Alexander Pushkin (1799 – 1837), the greatest Russian poet. –Editor's note

up, "Mamma, which Pushkin should I read – blue or brown?"

"Perhaps, read the brown," I answered. "Maybe you will find some of Poppa's notes, or the book will open upon the pages he most often read, and if you are lucky, you will capture what he thought about at the moment and what he felt. Any of my perceptions about Pushkin I can tell you myself."

And I myself, when there is a moment, glance into the brown ones. My blue Pushkins remain on the shelf, poor orphans ...

We had many art books too. We bought them, and we also often received them as gifts. This was long ago – during that period no one worried about prices. And there were people – students of Alik, my patients – who wanted to thank us somehow. And giving art books was the best expression of gratitude. We had no better pleasure than to curl up on the couch and gaze upon the reproductions together. But there was one special book.

"We will meet each other once in adagio-Vivaldi," wrote Akhmatova,[11] and Alik and I met under the paintings of Bosch. Looking at them, Alik fell into the depths of analyzing historical developments, spoke of creativity, human genius, and contemplated the mechanisms of discovery, attempting to reach

> "...for the very essence.
> In work, in searching for the path,
> In the heart's turmoil.
> For the essence of days gone by,
> For their causes,
> For foundations, for roots,
> For the core."

In these moments, I had my eyes wide open and could magically see how he was capable of

> " ...always grasping the thread
> Of fates and events,
> To live, to think, to feel, to love,

[11] Anna Akhmatova (1889 – 1966) was the heart and soul of St. Petersburg tradition of Russian poetry over the course of half a century. Akhmatova's work ranges from short lyric poems to epic cycles, such as *Requiem*, her tragic masterpiece on the Stalinist terror. The above line is from *Midnight Verses*. –Editor's note

To attain discoveries. "[12]

Together we penetrated the darkness of centuries, departed into the universe, and climbed into the heights of human understanding. And if father Aleksander (Men') was correct in preaching against physical love giving priority to platonic love, then these were the moments of our highest union...

* * *

Alik maintained warm contacts with his ex, Ira Karpova, his "one and only wife," as Natasha likes to emphasize.

After many years, already after Alik's death, Ira and I became friends; she and Natasha have a very tender relationship. When Natasha was twelve years old, Ira painted a portrait of her which now hangs in Natasha's home in France. Ira was one of the first people Natasha called to inform that she had given birth to a daughter...

While Alik was still alive, several times a year on certain prearranged dates, Ira would call him, and they would meet. At first, Alik would be careful in telling me about these *rendezvous*, but later, seeing that I considered them to be nothing more than completely natural, asked me, "And you react to this perfectly calmly?"

"And how should I react?," I asked bewildered. "After all you weren't born yesterday. You, like I, have a past which neither of us are intending to desert. You and Ira met and parted long before you and I met, so this is your story, and I have no connections to it. I have no pretensions on Ira's, nor anybody else's, place in your soul, just as my spot, no one else can occupy."

"It is great for you," sighed Alik, "you are not jealous."

But he was extremely jealous. He had no basis to be jealous, but jealousy is a character trait, not an inadequate reaction to external circumstances. Once, after consecutive and unannounced price hikes, I said with indignance, "What woman can permit herself to buy new stockings rather than fix her old pantyhose?! Only if she is going off on a date with a lover!"

---

[12]Excerpts from one of the late poems of Boris Pasternak, arguably the greatest Russian poet of the Soviet time. –Editor's note

"And so," Alik reacted with unexpected interest, "this is important information."

From that time onwards, handing over to me his salary, he invariably added, "But keep in mind that we do not have money for new stockings."

I have a long-time, "under the skin" friend – he and I have our own special relationship, our own topics for discussion, and during our rare meetings a third person is always too much. Both of us are night owls and can talk until morning. Alik, it appeared, understood everything, but every time he awaited my return from Lyonya and did not sleep. He never verbally objected, in any way, but simply waited for me and did not sleep, and I knew that after a sleepless night he could not work. Oh, well, the whole thing ended in that I began to return home from Lyonya's earlier.

Once Alik suggested that I invite to our home the poet N, an acquaintance of mine, about whose poetry I often talked. He received him with unusual friendliness. I did not at first understand what was behind his superpolite "*beau monde*" behavior. At one point in the conversation, the poet said that he wrote his diploma thesis on Pushkin.

"This is absolutely wonderful!" Alik pronounced with enthusiasm. "I have thought about Pushkin a lot and wished to talk about him with a professional. Here, for example, how should one understand his lines..."

In answer, the poet pronounced something rather shallow.

"Yes, yes, of course, but if one were to imagine..." continued Alik even more politely.

Again, something insignificant in response. And Alik continued further, "And what do you think, what did Pushkin think about..., how did he feel about..."

And in this way he continued until the end, speaking in a soft, piercing, well-intended tone. The poet was not even close to being a fool, but could not bear the depth of the analysis on which the proposed discussion was led and continued to answer in a rather helpless way. This was followed by an absolutely harmless question or comment made by Alik, then following up on the points made by the "professional." Before my eyes, a person was being destroyed. He was being destroyed politely, intelligently, and I would even say, sweetly. To my great regret, I cannot reconstruct verbatim this dialogue, and I do not wish to invent it. I can only convey my impression that this was not a machine-gun or tank attack, but a soft, almost delicate, crushing by a steamroller – I simply saw how a happy-go-lucky poet was turned into a grey, flat, and lifeless line.

It was extremely interesting to listen to Alik; his mind games brought me into a state of amazement. If he wanted to be a charmer, he could certainly do so, at least for me.

When the poet left, having happily rubbed his hands as if nothing had happened, Alik noted, "Pleasant evening we spent. It's only too bad that he sat around for so long."

"How can you behave yourself this way?!" I launched at him. "You totally destroyed a person!"

"You think so?" Alik asked me with a knowingly naive tone.

"Yes, I think so! And you know this yourself perfectly well. What did he do to you that you should persecute him in such a cruel, Jesuit way?! You are simply jealous. You should be ashamed of yourself! You are a scientist of world renown, but are jealous as the worst janitor would be!"

"And why does a scientist have fewer rights on his woman than a janitor!" continued Alik in the same naive manner.

"So now you are starting in on me! One victim is not enough for the evening?"

"So you do, indeed, think that I beat him? Well, this means that everything is in order, and now I can go to Vinnitskaya Street where mother has been waiting for me."

Need I add that the poet and I never saw or heard from each other again?

* * *

It would be correct to say that Alik could not stand vulgarity or low class culture. Such culture simply did not stick with him.

Once in his younger years, after an intense day of work, he went out for a walk and saw a theater poster which said "Sad Gvozdiki."

Having read *gvózdiki*[13] he thought that this had to be a charming comedy and decided to go and entertain himself. How disenchanted he became when the play turned out to be a melodrama "Sad Gvozdíki." With great difficulty, he waited until the intermission and then left.

Once, after Natasha's birth Alik, returned from his evening stroll earlier than usual and was almost green in color despite the walk. I became frightened that he was ill. It turned out that behind him two schoolgirls had been taking a stroll, and with his keen sense of hearing Alik unintentionally heard their conversation, which he related to me:

"I am an adult and you and I are close, but I cannot repeat even to you what these thirteen, fourteen-year-old teens spoke about – using which parameters and sexual criteria, they judged their young acquaintances..."

And this is despite the fact that Alik was far from prude, as he liked a sharp word, and believed that one can say anything, but the key is to say it properly. When I am in good mood, I like to entertain friends, relating to them "an episode out of life." Among my stories, there are piquant ones. Friends like them and sometimes ask for an encore. Listening, Alik always laughed, as if he heard it for the first time (or once in a while would forget?) and was clearly pleased...

In the 1970s, the Iron Curtain opened up a bit, and we began to find out about certain elements of life "outside." Once, having returned from work, I told of the following marvel: "One of our coworkers was vacationing in Bulgaria. Can you imagine, there are beaches where people go totally nude – men, women, children, all together."

"So, what? Sometimes during our camping trips, we also stripped bare."

"Camping is another story. You did that in extreme situations, as you had to, not just like that, while vacationing just for fun."

"I never took you for a sanctimonious person."

---

[13]Play on words: *gvózdiki* in Russian means "little nails," while *gvozdíki* "carnations." –Translator's note

"It has nothing to do with being a hypocrite. At my institute, I constantly see relatives who, due to a lack of service personnel, take care of their loved ones, as they wash, they clean, and they carry around chamber pots. Want it or not, they see the naked bodies of their relatives in all their intimate detail. This is an unfortunate necessity – illness, death. But under normal circumstances, nakedness should not be turned into normalcy, as it has its own function: in art – to evoke euphoria, in real life – desire."

"Yes?" In Alik's eyes there lit up devilish sparks and, after an expressive pause, "Perhaps you are right in some of this."

* * *

With his secretive nature and outward reserve, it was difficult to imagine what insane passions were boiling inside him.

"Ice and flame" – this was the only whole. He did not explain his actions, but the reason could be absolutely unexpected.

His hair became grey very early on, during the darkest period of his life – when he parted with Gelfand. Alik's papers were not published, and the significance of his work was not understood. He was embarrassed of his greyness, thinking it betrayed all of the suffering he had to live through. As much as I tried, I just could not convince him that grey hair on a man is beautiful and that women love it. But a complex is a complex...

At the end of the 1970s, there came from somewhere a fashion for the starvation diet. People fasted in different ways – they would go without food for one day a week, several days in a row, once a month, and in various other ways. Alik and I both laughed at this fad. But suddenly, Alik came home and declared that he had begun to fast. I should note, although I worked all of my life at the Institute of Surgery, I am convinced that without special medical reasons, a person's organism should not undergo such stresses.

"First off, before beginning to torture yourself, you could have discussed it with me. After all, your wife is a certified doctor."

"What is there to discuss – I knew it anyway. I knew you would be against it, so I locked myself in for two days at mom's apartment, and today is the third day that I have been fasting. So, come to grips with it."

"This is just ridiculous! Your health doesn't belong just to you! You should have a sense of responsibility for your family. You don't have the right to act so simplemindedly!"

"Don't worry. I am doing this under the supervision of a doctor-dietician. Later I'll have to come out the fast on a special diet, beginning with fresh squeezed juices. Here, I even bought a juice maker for this."

He fasted for ten days. Despite this, he kept working and, on the outside, behaved as if nothing was going on.

Coming out of hunger is an especially important moment and has to be done with products that are not polluted by chemicals. I could be against this as much as I wanted, but I ran to the market to buy organic vegetables and fruits and attempted to squeeze juices to serve them to Alik immediately after squeezing. Once in the kitchen, when I was silently preparing a portion of juice, my body language showing my indignation very explicitly, he said, "Don't be angry. I could not do it any other way. Over the course of one month Valera Nikolsky died and also my close friend in Kiev whom you did not know. Without an external distraction, I could not have overcome these losses."

I froze with a partially peeled carrot in my hand. Of course I knew that Alik was closely tied to Valera and understood that his death was extremely difficult – and, in addition to this, not a word from him! – but to such a level! Next to me was a person who was suffering so much that he almost died himself, but I absolutely did not imagine that this was actually happening!...

Alik's unusual reserve often led to major complications in his relationships with the outside world. Once he told me that in his youth he was interested in some young woman. They had a date set for one evening, but he decided to walk outside the city during the first half of the day. While walking he miscalculated the time, and at the last minute discovered that he was late for the commuter train, which would have brought him to town to the designated spot. So, he ran out to the train rails and then ran along the rails. The train caught up with him and the engineer wildly signaled at him, but Alik did not move from the railway, understanding that the commuter train would pass him by, he continued to run in front of the train along the rail track. The engineer was able to stop the train, and he jumped out of the train onto the rails and showered Alik with the rudest swearing, but, once he heard the explanation, he calmed down and let Alik ride with him. Arriving in Moscow, the elated and happy Alik started daydreaming on his way to the rendezvous, and as a result lost fifteen minutes and did not quite make it in time. The young woman was angry and said that the only reason she had waited for him was to explain to him that she would not accept such disrespect, and proudly walked away.

"And you did not tell her? You could have been hit by the train because of her!"

"What's the point, if the person does not understand ..."

I was twice the age of that girl, but I cannot say that that always helped. Misunderstandings could range from amusing to rather weighty ones. Once, during the first months of our acquaintance, Alik rather quickly ran off from our date, leaving me, to put it mildly, in complete confusion. "The events" developed in the following way. During our previous meeting, he asked why I always wore slacks.

"Do you have something against that?"

"No, but skirts and dresses are more feminine."

This comment was enough for me to immediately dig out from my closet my old boots with heels, since I wore more casual shoes with my pants. I urgently took them to get repaired, and, to our next date, I ran off wearing my spike boots and a skirt to my navel, thinking myself the apex of femininity in such clothing. At that time, we had decided to go to the Pushkin Museum[14] to view the portrait exhibition. And then, in the festive silence of the museum, I discovered that my newly repaired boots loudly squeaked. I already knew then that Alik physically could not stand some sounds. I felt disinterested with the pleasant portrait art, only thinking how I should move around the least, but, with the slightest of movement, the cursed boots simply shouted across the whole entire museum.

[14]The Pushkin Museum of Fine Arts in Moscow. – Editor's note

Feeling positively unhappy, I stood before a painting of an unknown artist on which was a full body image of a nine-months pregnant woman, in a black and red striped robe. She had green eyes half the size of her face, full of tears, and so much sorrow and anguish emanated from the entire painting, that I, somehow unwillingly, thought to myself aloud, "Such was the wife of Andrei Volkonskii..."

"The little princess? Why?" asked Alik.[15]

"She likewise was afraid of her belly, as she felt that it would bring her death." As soon as we left the museum, Alik quickly said his farewells. "It is them, those damned boots," I scolded myself, remaining alone.

He called that same evening. "You're absolutely correct, concerning the painting with the pregnant woman. Yet it was not the little princess but the wife of Prince Yusupov; she died from fever during childbirth. The painting was painted by one of the prince's court artists. I am calling just to express to you my admiration of your keen understanding of art."

"If Alik only knew out of what "garbage" my remark came!" I smirked to myself and said aloud:

"And how did you find out about the history of the painting?"

"I knew that today there would be a program about this exhibit, so I hurried home so mother could at least watch it on the television since she cannot go to the museum herself. The program turned out to be interesting, and I continued to watch it. And there I heard a discussion about this specific painting."

Well, then, all's well that ends well. But if I had not made my comments about the painting, if they had not discussed it on the program, if they..., if they..., I would never have found out that my poor boots were not, in any way, at fault. Would it not have been much easier if he would have simply explained what was going on?...

Once, when we lived together on Vinnitskaya Street, in an absolutely lighthearted way, I said, "I became a cook because of you."

The reaction was surprisingly sharp:

"And you believe that it is just a cook I need, and I don't deserve

---

[15] Characters from Leo Tolstoy's novel *War and Peace*. – Editor's note

anything better?"

Well, here I had much to say: explaining to him, based on popular understanding, that everything depends on the man – if he feels himself to be deserving, then he will consider his wife a queen even if she is in life, in reality, a cook. But a weak king will treat his queen/wife as a cook or, even worse, as a court dog. In general, we played out an old anecdote "mother, he called me a bitch." Many years later, long after Alik's death, Louisa Kirillova told me that she had always liked Alik, and feeling sorry for him for his unkemptness, once she decided to marry him to her maid. Alik bridled at this suggestion.

If one could know and understand!.. How many misunderstandings would it have been possible to avoid? And how many of them remained undetected ...

But sometimes at the edge of a disaster he betrayed his own rules. Once when Natasha and I were moving into our own apartment on Kedrov Street, there occurred a chain of misunderstandings and accidental coincidences, on which it is pointless to reflect further, but at that moment, I became wildly angry, grabbed Alik's things, and carried them to Vinnitskaya Street. Alik was at the university, and I left a note in two words: "Return keys." I did not return his phone calls or letters, just returning the latter unopened. Then he called my girlfriend Alka – with whom he had rather distant relations – passionately explaining the situation to her for over half an hour. The very fact that Alik, in all of his reserve, tried to explain something, which was weighing on his mind, was so huge that I immediately called him and only said, "Come back." We never talked about this incident again – why should we? We knew what we needed to know.

* * *

If Chekhov was correct in saying that brevity is akin to talent, then this was true about Alik. I remember one case in the late 1970s. Winds of change were sweeping the times. A series of sharp articles appeared in the press. Someone once brought a fresh newspaper to work with something sensational in it. The paper passed from hand to hand, but I was called into the operation room, and later I had to attend a patient in recovery. I only had time to call Alik and ask him to pick up Natasha from kindergarten. I returned home late that evening and said, "Do you know that there is an interesting article in *Izvestiya* ?"

"I do. Here it is. I put it aside for you."

"I have absolutely no time to read it. I have to urgently cook dinner for you two and then do the wash. Just tell me what it's about."

He was silent at first and then said a few phrases.

"And is that all?"

"That's all."

"But Alik, there is a whole large piece! What else is there?"

He thought a while and then said: "But I don't think there is more there."

But then I opened my mouth, "Of course, if it were Nikita, or someone else from amongst your smart mathematicians, you would have discussed this for hours, digesting every paragraph, but with me it is possible to get away with two sentences!"

As in Pushkin's fairy-tale "the golden fish did not say anything, only waved its tail and swam out to the open sea ..." Alik looked at me silently, shrugged his shoulders, and vanished into the next room to sit at his desk.

Having done what was absolutely necessary and could not be postponed till tomorrow, and having put Natasha to sleep, I finally got to the article. First I looked over it then carefully read it from the first line to the last. But I didn't find anything that Alik had not noted! (I absolutely do not remember what was in it. Apparently the story was one-day news.)

There was nothing to do but go and apologize...

Brevity, apparently, also became Natasha's inherited character. Once, during the first months of her doctoral work, Natasha complained that her advisors expressed dissatisfaction. They proposed that she and another student write an intermediary report about the work done to date. They went to the advisors together, who on accepting the papers, both noted, "It is immediately clear how much the young man has done; instead of bringing the required report written in three pages, he brought one in four and a half. And what about you, Your Highness, you only worked to write on one and a half pages?"

Natasha explained with bitterness, "The second doctoral student gave me his report to read: he had familiarized himself with one article, decided to repeat the experiment described in it, and for this, took the alcohol, heated it above a beaker, then took into the pipette the solution, emitted out of it several extra drops, so that the solution would be exactly to the required mark, then released it into the beaker; and then, in just the same way, did the same with three other fluids; began to heat the mixture, and etc., then did not achieve the results described in the article. That's all. And I wrote that I read fifteen articles, determined that the questions posed before me are similar to those being worked on in other laboratories around the world and are being interpreted in three different directions. In all these three directions, I carried out several preliminary experiments, picked the one that seemed to be the most promising to me, especially after a small alteration." (Eventually, she received an invention certificate for this "small alteration").

At the moment when Natasha was speaking a thought passed through my head, derived from Pushkin. Not all of Alik is dust. His spirit in Natasha will survive, safe from the worm...[16]

* * *

Alik believed that a person who possesses a talent is responsible for it. He scolded me, "You have various and exceptional abilities. How can you permit yourself to neglect them?"

"Well, you will write on my tombstone *the tomb of the unknown wasted talents*," I laughed it off.

Alik responded angrily and without humor, "You boast that you are not ambitious, but the reality is that this is your biggest fault."

He viewed talent as something like public domain, which providence handed over to individuals, like precious grain handed over to

[16]Echoing a line from A. Pushkin's *Exegi Monumentum*, translated by Vladimir Nabokov. A more literal translation is: "Alik's soul in Natasha will outlive mortal dust and will escape decay..." –Editor's note

them for preservation, obliging them to sow these seeds and foster their growth, sacrificing their own strength, even their own life. And as a gardener shows off the fruits of his labor with pride, Alik felt it necessary for his scientific works to be valued. He was particularly ambitious and suffered from the vacuum in which he existed almost all of his creative life.

During the course of the past quarter century, so much has changed that even we, participants of this recent past, forget some of the details. While preparing these notes, I opened Alik's archives and stumbled upon some papers which simply scream injustice about how sadly the "powers of restraint" behaved with him. In one statement to the editorial board of the journal *Izvestiya Akademii Nauk, Series Mathematics,* Alik writes that the editors had suggested that the article which he submitted to them be divided into several parts, since it was too long, and that it should be printed in its parts in separate but consecutive issues of the journal. When he divided the article into three parts, the first went to press six months after it was submitted; the second did not come out in the following issue as promised but nine (!) months later; and as for the last one, he never received an answer about when it would see the light of day. I present verbatim the end of his letter:

"The issues considered in the article have been intensely studied recently in physics, as well as in mathematical literature. I have no doubt that if my article is not published relatively soon, the findings reported therein will be repeated by somebody else in the USSR or overseas. I am an ordinary mathematician but do not consider that this circumstance can serve as the basis for improper treatment of my work. To my mind the proper treatment in this case would be as follows: the Editorial Board determines the fate of the article in a relatively short time, say, three to four months. And if the decision is affirmative the time from the moment when the article was received by the editors to its publication should not exceed the average for the journal..."

And such examples, unfortunately, are many. Once Alik told me, "I should live fifteen years in the future. Only in such a mode I am able to exist in science."

When I asked why he had complications with the publication of his work, he answered that the country had two mathematical journals: one of which was anti-Semitic and, thus, did not publish the works of Jewish authors, while the other journal was headed by Gelfand with whom he had parted ways. I do not know the details of why he quarreled with Gelfand. Alik refused to talk about this topic. I think that both secretly regretted this extremely unfortunate event. My suppositions in relation to Gelfand are based on a time when Albert and Lucy Schwartz invited us and Natasha, still a school girl at that time, to Bures-sur-Yvette near Paris where Gelfand, who was also present, expressed the desire to meet with the daughter of Berezin; he conversed with her for about an hour, after which he said, "You have the same manner of thinking as your father. You can be a mathematician." I interpret Gelfand's gesture as a tribute to the memory of Alik. (But Natasha chose chemistry, "a field of science that is in the middle between father's mathematics and mother's medicine," as she semi-jokingly explained her decision. It saddens me that during Natasha's school years there was no one around to help her feel the beauty of mathematics.)

Alik's fame came to him through physics, and this did not come about quickly. Once I asked him whether he had ever attempted to converse with Landau. His answer was, "Once, but it didn't turn out too well; it is difficult for me to talk about it."

Here I would like to say a kind word about Igor Kobzarev.[17] He was amongst the first physicists who understood the significance of Alik's work and did much to bring it to the attention of theoretical physicists.

Alik led a seminar which was attended by many people from the Physics Center in Chernogolovka.[18] For Alik, this seminar was something sacred. Nothing, even for example the "trivial" issue of my dissertation defense, could bring him to cancel his seminar. He greatly

[17]Igor Kobzarev (1932–1991), a theoretical physicist and a member of the ITEP Theory Department. –Editor's note

[18]Chernogolovka is a small town located 50 kilometers (30 miles) northwest of Moscow. That's where the Landau Institute for Theoretical Physics headquarters were located, along with other Soviet research institutions founded after 1956. –Editor's note

lacked personal contact with the leading scientists from abroad. Picking through his papers, I found an enormous file full of invitations from all over the world to work overseas. Amongst these invitations were a number of rather weighty and prestigious ones which gave honor not only to Alik but to Soviet science.

... One forgets now how much humiliation one had to endure while filling out the paperwork in order to travel abroad – all of these professional committees, party committees, regional committees... A mountain of forms had to be filled out, with nonsensical questions, requiring answers "did not" and "did not associate." It would be good to write "member of the Party.[19]" But Alik did not want to join the Party, no privileges could tempt him to do so. And, thus, after all the humiliations and an enormous waste of time, the documents would be submitted to OVIR,[20] who would refuse him an exit visa, or simply never reply... Among the examples, I will mention only two of the most flagrant.

Alik's trip to Poland was killed already at the level of the University under the pretext that at that moment he had three graduate students and, seemingly, they could not be left without an advisor, despite the fact that Alik's stay in Vroclav would have been from May to December, an absence of one semester maximum. When he was invited to CERN[21] for a year and a half (he was offered to pick the length of his stay in Switzerland), the University could not officially refuse him this honorable invitation. After all the documents were sent off to OVIR, the University revoked Alik's references which had been issued by the very same University! At this unusual development, even the seasoned workers of OVIR were surprised, as they told Alik that this was the first time that they had encountered such a thing! Recreating the atmosphere of those times, one has to remember that there was no Internet, thus, there was not even this

[19]The Communist Party of the Soviet Union, the only one that existed in the USSR. –Translator's note

[20]Department of Visas and Permits, a section of the Ministry of Interior Affairs which was in charge of exit visas. –Editor's note

[21]Centre Européenne pour la Recherche Nucléaire, European Organization for Nuclear Research. –Translator's note

avenue for connecting to the outside world.

Alik suffocated in the nation of socialism whose face was so far from being human...

It was critical for him to know that someone expressed an interest in his work. I will relate one such episode. Once, late in the evening, he told me that the day after tomorrow he would have to finish a major project and, therefore, he would not answer the phone, asking me to pick it up and let him know who was calling so that he could determine whether to talk to that person or not. In the morning, Vitya Maslov called. Alik gave me a sign that he would take the receiver, and... talked for forty minutes! Thereafter, he sat without getting up from his desk for the rest of the day. I had to reheat his lunch twice. When we finally sat down at the kitchen table, Maslov rang again. Grabbing a piece of bread, Alik ran off to the other room and talked away for about an hour. To all of my gestures signifying that lunch was getting cold, he waved me off and signaled that E.A. and I should not wait for him. After eating, I left his food on the table and, angrily, went off to play with Natasha. Having sipped the cold soup, very pleased with himself, Alik appeared at the door and, as a sort of apology, explained that Vitya had become very interested in his work and decided to dedicate his whole day to studying it. In the morning, Alik sort of delivered to Vitya an introductory lecture, and, at midday, Vitya called to tell him that he had reached the middle of the text.

"In the evening, he will call me again when he is finished," added Alik. He did not go for his evening stroll – he waited for the call. Maslov called late, and again they chatted for a long time. That night Alik slept well, wheezing away in his sleep with satisfaction. Apparently Maslov's attention was a good substitute for the missed evening stroll.

...Well, my gratitude to Vitya that he was interested in Alik's work while he was still alive...

* * *

At one time, a young, beautiful woman often came to see us. She eagerly exchanged pleasantries with me at the entrance, gave

Natasha some sort of toy, and then she and Alik would disappear into a room, closing the door tightly behind them, where they stayed for long periods of time. I had the right to clean, do the laundry, cook, go out for walks with Natasha, or put her to bed – no one bothered me with pulling behind myself all of the household chores. Finally, the door of the room would open, and the tired but satisfied pair appeared in the corridor. The woman would quickly say her farewells and leave while Alik sat down to dinner. Once after such a visitation, I said:

"However you look at it, you have an ideal wife."

"The most important thing is that you think so," smirked Alik.

"Yes, ideal," I maintained. "Maybe for a normal man I am not at all satisfactory, but for you, ideal!"

"I repeat, the most important thing is that you think so."

"But tell me, what kind of a wife would put up with her husband shutting himself up in a room for many hours with another woman?!"

This time, Alik took the time to explain. "The three of us – Renata Kalosh, Ogievitsky, and I – received an invitation to an international conference (congress? – I do not recall). This Ogievitsky and I, as Jews, naturally, were discounted right away; but Renata – the daughter of a Hungarian communist who was shot under Stalin, but now rehabilitated with honors – she is permitted to go. This Renata is an insightful specialist in her field, but she lacks a general scientific overview. This is the first time she is traveling abroad. She is very nervous. And I, so to speak, am trying to give her a crash course on the present state of scientific research."

"Out of solidarity with you, she should have refused to go on this trip."

"No, Renata is in no way a traitor. She understands our work and will present all three of our papers."

"Still, this is not the same thing. She will not be able to present the papers as the authors would."

"Well, as far as I'm concerned, she will do it better than me. Renata had a good idea. In the beginning of her presentation she will write down on the board two formulas, one next to the other: what the situation was before and then my proposition. I would never have thought of this myself, " added Alik with obvious disappointment.

I attempted to comfort him. "Everyone's his own: Renata – she is an artillery attack, but you – penetrating radiation."

* * *

Well, well, but a penetrating radiation he was. This was especially apparent from the time when he was no longer with us. Once a year on the 25th of May, for ten years before we left for France, I gathered all the people together who wanted to remember Alik. And with each year, there were larger groups gathering. And even during the old times, people I did not know at all came over to me simply to tell me how much they missed Alik...

Once Volodya Molchanov came and asked for a photograph.

"Do you know," he said to me, "we saw each other rarely and spoke on the phone or communicated by the post not very often. But now that he is no longer with us, I feel the need to at least see his face sometimes."

I gave Volodya the last, best, photograph of Alik. He is in Novosibirsk, by the blackboard, speaking with passion about one of his works which, finally, had captured the interest of the scientific community. Volodya had the photograph enlarged and made many copies which, at that time, was not at all as simple as it is today. I gave them away to close friends; this photograph is also on the jacket of the book which came out after Alik's passing – *Introduction to Algebra and Analysis with Anticommuting Variables.*[22]

Natasha was twelve years old. We were spending the summer in Mikhalevo in the "House of Scientists." Once the Mikhalevo society became very excited – a young theoretical physicist and doctor of science, recently transferred from Tomsk to Moscow, a bachelor, appeared there – wow, what a catch! He was placed at the directorial table where, thanks to my friendly relations with the directress, Natasha and I also sat. All of the Mikhalevo available beauties sat

[22]F. Berezin, *Introduction to Algebra and Analysis with Anticommuting Variables*, ed. V.P. Palamodov, (Moscow State University Press, 1983); English translation: F. A. Berezin, *Introduction to Superanalysis*, ed. A. A. Kirillov (D. Reidel Publishing Co., Dordrecht, 1987).

there, as well. After lunch, Ella said to me, "However you look at it, your daughter has an outstanding appearance. So many beautiful women were sitting at the table, but during the course of the entire meal, this young scientist did not shift his gaze from Natasha."

"Well, " I smirked, "he does not look much like the hero of Nabokov's Lolita. Most likely he had crossed paths with Alik within some sort of a mathematical circle, and since Natasha looks very much like her father, an association has formed now in the memory of this young man."

"The things that you can invent simply do not fit through any barn doors." Ella shrugged her shoulders.

During dinner, she demonstratively asked with an indifferent tone, "Tell me, Dima, does the name Felix Aleksandrovich Berezin mean anything for you?"

"Well, of course. I continuously use the findings in his book on the secondary quantization formalism and supermathematics in my work. The discussions, which I seldom had the opportunity to lead with him, always permitted me to find solutions to all my problems. Besides that, he was the opponent [23] for my doctoral dissertation. Without him, this defense would not have been possible. For me, this is a person and scientist without equal. In general, I am a person who is not jealous, probably because I don't have many complexes. But I always envied my Muscovite friends Tiutin and Voronov who had constant opportunities to attend Felix Aleksandrovich's seminars and discuss various scientific issues with him."

---

[23] The academic hierarchy in Russia follows the German rather than the Anglo-American pattern. An equivalent of the PhD in the US is the so-called *candidate* degree. The highest academic degree, doctoral (DSc), is analogous to the German *Habilitation*. Thesis defense in the Soviet Union was a lengthy and complicated procedure. Dissertations ready for defense were sent to the so-called *official opponents* – experts from other institutions – who were supposed to study them carefully and then present their critical evaluations on the day of defense. Thorough debate was supposed to follow, and only after that could the Scientific Council approve (or decline) the dissertation under consideration. In the case of approval, dissertations were then sent to Moscow to the Supreme Dissertation Council (VAC) for further consideration. VAC had the ultimate say: it could either confirm the degree sought, or veto the positive decisions of the local Scientific Councils. – Editor's note.

"Well then, please meet his widow and daughter."

"Here I am struggling – how do I know this girl?" exclaimed Dima. "I just could not understand of whom she reminded me. But as it happens, I saw her when she was still very little when I came to Felix Aleksandrovich's home in order to discuss my dissertation."

Laughing, I said to Ella, "As you see, Bohr's suggestion that a theory is correct if it is sufficiently crazy is valid not only for theoretical physics."

After dinner, Dima came up to me and said, "May I go on a stroll with you? I very much want to talk about Felix Aleksandrovich."

Thus, unintentionally, on this evening Natasha and I surpassed all of the Mikhalevo beauties.

Dima Gitman became a major friend at our home. He described to me how he had to defend his doctoral dissertation in Novosibirsk at the Institute of Nuclear Physics. This defense was hindered by a number of "nonscientific reasons." Then Efim Samoilovich Fradkin informed Alik of the situation and asked him to be the opponent at this defense. It was believed that no guardian of racial purity could withstand such "heavy artillery" as the Berezin-opponent. Dima was shaken and endlessly thankful to Alik for consenting to be the opponent. It goes without saying that Alik fully probed into the essence of the work. I recall that Alik flew to Novosibirsk and was very satisfied with the results of his trip; as usual, he did not enlighten me with any details. Dima told me that their foresight was not without reason – one of the most influential members of the Scientific Council began his offensive against his dissertation on abstract problems of the quantum field theory with such questions as, "And how from all this can one get practical applications?" At the same time, Alik's presentation in support of the work was so informal, succinct, and interesting from a scientific point of view, and, at the same time, diplomatic in relation to the "patriots," that he immediately and sharply changed the atmosphere in the auditorium; the vote was unanimous.

Dima gave Natasha and me an audio cassette with a recording of this presentation. As far as I know, this is the only opportunity to hear Alik's voice which, incidentally, was very beautiful – such a low baritone...

There are numerous examples of "the consequential effects" that ensued due to Alik's interference.

In 1972, there was an international conference in Moscow. Alik said that he must meet with one of the foreign scientists who came to this meeting. At the end of the conference, I asked, "Did you talk to the person you wished?"

"Well, how should I say..." Alik shrugged his shoulders indeterminably.

"Tell me what happened."

"Well, I asked my question during his presentation..."

"And?"

"He began to answer, then stopped and became quiet; as silence prevailed – a rather impressive silence – he then said that the question was very interesting, and that he couldn't answer it on the spot, but after going home, he would certainly think about it. And when he thinks about it," Alik said with his characteristic devilish smirk, "then he will understand that his entire construction falls apart..."

Once one winter when I was heavily pregnant, Alik decided to take me out of the city to breathe some fresh air. We rode in the commuter train and stood near the entrance. At some small station a skier popped in. Alik and the skier greeted each other and, without even thinking of introducing us, Alik said to him with great energy: "I was just thinking about your work..." And then followed the typical abracadabra.

... When Alik spoke to mathematicians, it seemed to me that if they conversed on "normal things" in Esperanto, I would have understood more than that Russian mathematical language they used. Still, I had a feeling that I understood their hierarchy of power. Although Alik spoke quietly and never raised his voice, he almost always was in the leadership position. And this is understandable – most of the time I heard his discussions with his students. But even when his colleagues from the university came to see us, Alik was in the leading role most often. My words are supported by at least the fact that the letter to the Dean written about the situation at Mekh-Mat (see Part III of this book) was discussed at our home by the colleagues whom Alik gathered. Initially, this letter was planned to have been from the collective, but later it was decided that only Alik would sign it. For this letter, the university administration, as was often the case, had their revenge...

And here, near the entrance to the commuter train, I saw how Alik's words made an impression on his acquaintance, as if he received a hit to his gut. He stood back, began to withdraw, but Alik, feeling satisfied by such an effect – not allowing the "rival" to come to his senses – continued the offensive. The skier took a deep breath, concentrated, and began a counteroffensive. And here the time came for Alik to withdraw... And the battle of the minds continued. In front of me was not a game of checkers or chess – this was not boxing or fencing, but an unknown Martian battle of Titans. Alik had encountered a worthy opponent, and I stood frozen, watching this duel as a die-hard fan cannot tear his eyes from the ring. Alik, however, without any doubt, was actually showing off in front of me. Because of this, an immediate payback came – he answered too quickly to some sort of contention from his opponent and, because of that, did not think things through enough. And here there was no mercy for him. I felt that I was distracting Alik's concentration, and so I walked off into the car. When he came after me, Alik answered the question of "Who won?" with, "We did not finish talking since the train came to his station. But this is not important, since the most important thing is that he will have a chance to think about it."

"And who was it?"

"Arnold."[24]

* * *

Concerning the relations Alik had with his students, unquestionably, they can speak about it better than I. I will only mention a few episodes.

A near total fiasco occurred with his first student. This was a young American who, upon first arriving to Moscow, was happy,

[24]Vladimir Igorevich Arnold, an outstanding Soviet mathematician best known for the Kolmogorov–Arnold–Moser theorem regarding the stability of integrable Hamiltonian systems (KAM theorem). – Editor's note

energetic, and jumping with enthusiasm to work. But after a little while the young man began to somehow wither away, become sickly, and ceased to express any interest to work on proposed projects. Alik did not understand immediately that the reason was not in the American's disenchantment with mathematics but that he was fading from hunger! Yes, in the center of Moscow, living at the university dormitories, the young man was starving. And the thing is that this had nothing to do with the absence of money, as he was from a very well-off family. Simply, it is that every time he came to the student dining hall, he encountered a long queue and, therefore, decided to come back later. But people stood in this line until closing. As for Soviet stores, aside from the queues, the American was scared off by the unappetizing appearance of the offered goods. Later, when all of this came to be known, measures were put into place (if I am not mistaken, the American was permitted access to the professors' dining hall), and everything ended with an excellent finish for his term of study.

But Alik remembered this lesson for life and, for this reason, was extraordinarily attentive to his students. Numerous times he directed them to see me for medical consultations. As a rule, the young people did not have any organic illnesses – to a large or small degree, they suffered from what the French have so precisely christened as *mal de livre*, i.e. dissatisfaction with self and loss of one's place in the surrounding world. They did not need a medical healer but a psychotherapist. But, since in those days such a profession practically did not exist in the USSR, I had to play the role of "self-taught Freud." Usually as soon as the first positive results began to appear in their work, my patients magically recovered. And the credit for this goes not to me, but to advisor.

Very often, Alik helped his students not only in purely scientific explorations. Thus, for instance, one of them defected from the Soviet Union as he crossed the Finnish border in the region of Kolsky Peninsula on skis. Alik knew about this desperate act and was very worried until he received news, via underground channels, that the operation had been a success. When the young man was beyond the border, he was in need of a recommendation letter. And, when it was possible, Alik sent along such a letter. In the situation that reigned in the country at that point, this was a truly brave act.

Once a foreign graduate student, an Egyptian Arab, came to Alik and said that he wanted to write his dissertation under Alik's direction. Alik was puzzled since he was not sure whether the Egyptian knew that he was choosing a Jew for his academic advisor. He attempted to convince him to appeal to other professors at the university under the pretext that the young man had weak preparation in his specialty and, therefore, he would not be able to finish his program in the standard three year period. But the graduate student answered that he wanted to work specifically under his guidance and was prepared to dedicate as much time as necessary; that he could choose any institution of higher learning but chose Moscow State because he had become acquainted with the works of Berezin and it was those areas of research that were of interest to him. Then Alik asked him a direct question, whether the young man knew that his advisor, in such a case, would be a Jew.

"I know everything about you," answered the young Arab, "even that you recently had a daughter."

In the course of the first two years, they spoke only of mathematics. Later, gradually, they began to touch upon various social and political questions. The Egyptian related that, in his country, there was no total anti-Semitism, that Sadat had signed a peace agreement with Israel not from the position "one against all else," but rather with the support of a specific strata of Egyptian *intelligentsia.*

When, at the end of his second year of studies as a graduate student, he was departing for his summer vacation and, unlike his advisor who could not go abroad, planned to tour around Europe, he asked what present from abroad would bring Alik the most pleasure. Alik answered that he would like to have a current edition of the Bible, since in our country at that time, the Soviet authorities did not permit publication of the Bible... And here is one of the paradoxes of the life we lived then – I was in my mid-thirties, and this was the first time I could hold a Bible in my hands, which an Egyptian Arab brought us from England!...

After the defense of his dissertation, the young man, who at that time had already become the father of a family, said that, aside from the usual banquet, he and his wife wished to come to our home and celebrate the occasion in a closed circle (they lived at the university dorms and could not invite us to their place). Furthermore, the wife of the young candidate of sciences insisted that she prepare all that was necessary for the celebratory table. They brought with them not only victuals but their kitchen wares. When I returned from work, the apartment was full of the delicious aroma of pilaf, the table was laden with oriental dishes, and on the cupboard oriental desserts awaited their proper hour. We had a splendid evening!

Dima Leites once told me, "Felix Aleksandrovich towers over us, mere mortals, but often, after conversing with him, I begin to feel better about myself."

"There is a general rule for a teacher and doctor – after communicating with them, even the most unfortunate student (this, of course, does not relate to you) or helplessly ill patient, at least for some time, should feel better," I noted.

Alik never permitted himself to have a condescending tone. He had an amazing ability to push the person he was talking to, to think. (I attempted to express this character trait in the dialogues I had with him that I have recounted here – I do not know whether I have succeeded).

As a rule, after finishing collaborative works with students or younger colleagues, Alik maintained contact with them and warm relations continued. I know of only one case of a total and absolute break with a young mathematician, who, in the winter of 1975-76, spent long hours at our place on Vinnitskaya Street, working with Alik, and, after his immigration to the USA, decided to publish a coauthored work there under his name, without the coauthor; he informed Alik about this with a letter which he sent on the eve of his immigration out of the USSR. I cannot express what Alik was going through when he received this letter! I was afraid of a heart attack due to nerves. One has to imagine the circumstances of those times: Alik was denied the ability to travel abroad, and his works were either not published or published with endless delays. Aside from the nature of the situation, Alik was also disturbed by the manner in which this was done. Probably only Natasha's birth a week after this event saved Alik. What is more, the young man had no need to behave in such a manner, since Alik considered him a rather gifted mathematician and foretold that he would have a brilliant career if the circumstances were right, something that did indeed happen.

I do not want to end my stories about his students and young coworkers on a negative note and would end with pleasure with several more words.

Alik met Misha Shubin and Grisha Litvinov at a mathematical school near Baku. The three of them were placed in one room, and, after several evening conversations before going to sleep, Alik suddenly proposed to Misha and Grisha that they write a book together. This work was never finished, but the relationships that sprouted, one may say absolutely unexpectedly, always remained very deep.

After Alik's passing, Grisha came to our home and offered to teach Natasha mathematics. This did not last long and it was apparent why: the psychological foundation was incorrect, since both Grisha and Natasha were looking for Alik in each other. But this pain Grisha had, I recall even now...

Misha did quite a lot for Alik. After Alik's death, he arranged the translation of their book on the Schrödinger equation which had been previously released in Russian by Moscow University.[25] Due to Shubin's efforts and those of his colleagues, several collections of scientific works were prepared and published and dedicated to Alik's memory. Without the gentle but determined pressure of Misha and Albert Schwarz, this memoir of mine which you read now would never have been written.

When Alik first invited me to his place, there was a huge celebration. He gathered his friends to celebrate the release of *Lectures on Statistical Physics*.[26] On the table stood absolutely inedible wonders

---

[25]F. A. Berezin and M. A. Shubin, *Lectures on Quantum Mechanics*, (Moscow State University Press, 1972), Expanded version was published in Russian in 1983 under the title *Schrödinger Equation*. English translation: F. A. Berezin and M. A. Shubin, *Schrödinger Equation*, (Kluwer, 1991). –Editor's note

[26]F. A. Berezin, Lectures on Statistical Physics, in Russian (Moscow State University Press, 1972). In 2002 the Second Edition was published by the Izhevsk Institute of Computer Research. –Editor's note

purchased in the Soviet stores, amidst which stood out pleasantly a very delicious beet salad (*vinegret*) and French salad (*olivye*) which had obviously been prepared by Masha Shubin. Masha herself was not there – at that time she was still a breastfeeding mother who, after preparing everything before the guests' arrival, went off to be with the baby.

Later, when Natasha was born, the Shubins passed on children's strollers, clothing, and toys to us. Alik and I were old parents, and the children of his students turned out to be our daughter's peers. Masha, as a more experienced mother, gave me wise advice on how to take care of Natasha. Our daughters, after they grew up, became friends. Fate has now scattered them to different countries and continents, but they maintain good relations. But in our Moscow life all of the birthdays were celebrated together. Once, after Alik's passing, we were invited to Galya's birthday – Shubin's youngest daughter. At the height of the celebration, there was a power outage. Masha sighed that she would be unable to feed the children – all sorts of *bliny* (crepes/pancakes) were planned, but the stove was electric. Candles were lit. The children were handed out dry cookies and candy. Masha and I conversed in low tones in the kitchen. Misha was telling something to the children in the other room as they sat there quietly chewing the cookies and playing with the wrappers from the candy. It was cozy, warm, and quiet. And it seemed to me that the power outage was not accidental, that Alik's soul was there, with us, hovering around the apartment, hiding in the trembling shadow from the flickering candlelight...

* * *

But that was later, later... Now, at this moment, the plane was going to land. We were arriving in Krasnoyarsk.

Krasnoyarsk – Magadan

Krasnoyarsk, Siberia... One imagines something huge, powerful. But the airport turned out to be small and cozy, something like an Aeroflot village depot. The airplane taxied almost to the arrival hall, where we had the chance to disembark and stretch a bit after the lengthy journey. Through the bulletproof glass, which separated the passengers from the airstrip, we could see a refueling truck driving up to the plane. At the same time, the baggage began to be unloaded in the most basic way – the door was opened and baggage was tossed down to the ground. The young tourists, laughing and joking, began to drag their kayaks out of the general pile of items. I watched them and mentally saw them off with the words of the song which they had sung during their journey:

"It's time for goodbye, my boys ... it's time for goodbye, my girls. You must try, you must try to return..."[27]

The young people were going camping, while I was traveling further and thinking how I should tell another girl that she would never see her father again...

* * *

From the moment we started to get closer, Alik wanted a child, but I did not. I had no maternal instinct. Alik was disappointed and upset by this. But I was not ready to give in.

"But you would not change your everyday priorities," said I. "You would continue to work 36 hours a day, and all the work with the child would rest on me. Of course, I am not as important of a scientist as you are, but I do love my work, and I am satisfied with the lifestyle I lead presently. Furthermore, we are already old – where were you before? You should have had children with the women who came before me."

"If I were woman, I would have had not just children but grandchildren by now!" Alik once replied to one of my tirades with an excitement unusual for him.

Our days of youth were far behind. I had turned thirty-six years old, and Alik was forty-three. This was not the first year we were together ... and I relaxed too much and was off-guard. But nature,

[27]From an early song of Bulat Okudzhava, the favorite *bard* of the Soviet *intelligentsia* in the 1960s, '70s and '80s. Translated by Tanya Jean Wolfson –Editor's note

as is known, does not tolerate when one neglects to take care of oneself...

During the first half of August 1975, we spent time together at a village near Dubna; then Alik remained there to work while I left to go to Bulgaria with a tour group. When I returned, everybody noted that I had never looked that good before. I felt myself somehow rejuvenated. I felt that some sort of joyfulness had descended on me. I liked everything, and I was pleased with everything. I did not notice any disturbance in my physiological functions, believing that this state was brought on by the changes in climatic conditions. Among the first to notice that something unusual was happening to me was our friend Lyonya. We were strolling along with him at Kolomenskoe, and Lyonya inquired, "What's going on? Where is your sharp wit, where is the nonstandard thinking? What are you doing with me today, feeding me sweet pleasantries instead of a substantive conversation?!"

"Well, listen," I answered, "you always scold me for being critical of everything. For the first time in my life I am satisfied with everything and, once again, this is bad for you."

"Well, this is simply not you!" Lyonya continued to be amazed and, stepping two strides away so as to look at me better, questioningly asked, "Perhaps you are pregnant?"

"Pregnancy or menopause – that is the question," I laughed.

At that time, the answer to this eternal question was an absolute "to be" – a new life was born inside of me. A feeling of absolute harmony with the outside world penetrated my entire being. At that time, I said that I had discovered a remedy from all conflicts and wars – the entire planet should be pregnant. It is only too bad that any pregnancy eventually ends.

From that moment when I was with child, Alik began to try to convince me to move in with them on Vinnitskaya Street. I did not want to move, but succumbed to the arguments – this way or that way, I felt good about everything. I loved everybody, even E.A.; and Alik expressed a mindless hope that we would all get along. Indeed, at that time, nothing irritated me, and I only had one determination – that life would be just as beautiful for everybody as it was for me. Lyonya proclaimed, "Remember 1976! This is the year of your wisdom." As if being wise is that simple: just a bit of a sober thought plus a whole lot of kindness and patience... The only question is, where does one get this patience for their entire life?

Looking back at that period, I think that it was not only my happiest time, but also the most significant in my life. What if I had been sent to this earth to bear Alik's child? Peace and satisfaction settled in my soul, in my subconscious; there was a feeling of fulfillment of fate. And it seemed that everything was against it: Alik and I were so different, no longer young, not intending to live together, and I did not actively wish to have children. But everything turned out to be the opposite. It seemed as if someone, somewhere had decided everything for us and led us along a path, which we did not choose ourselves, with a firm hand...

From the moment of Natasha's birth, she resembled her father uncannily. While still in the maternity ward, I wrote (at that time fathers were not permitted at births, as they are now), "If you had decided to reproduce by way of spores (cloning was not known then), then why am I lying here on this post-birth bed? You should have given birth yourself."

This resemblance between Natasha and Alik was retained forever, in character as in outward appearance – there is nothing there from me. I joked, "If men could give birth, then I would know for sure that you cheated on me."

From the very beginning, there existed some special connection. If Natasha cried, Alik knew better than I why: Does she want to eat, does she need to be changed, or does her stomach ache. She quieted down more quickly in his arms. He knew whether she was warm or cold, which shirts and footsy pajamas needed to be put on her. My mother was amazed, "You are a woman and a doctor, so why do you constantly consult with Alik how to take care of the child?"

"Because he understands better what she needs."

Once when Natasha was about a year old, on entering the room, I found the following picture: Natasha was standing in her crib, and Alik was standing still next to her. Father and daughter were looking at each other attentively, eye-to-eye.

"I just found an answer to the question about which I have been thinking for so long, and when I looked at Natasha, she understood me," said Alik. "You did not, but she understood."

His words did not provoke laughter in me: God knows them, what they understood about each other; they had a special plane to which I had no access.

Alik had his own way of preparing for the birth of the child: he read books about caring for and bringing up children. He had the idea to teach Natasha how to swim in infancy. So, at four months old, Natasha, with obvious pleasure, not only splashed in the water during bathing, but also "covered" the entire length of the bathtub while swimming. (In the aftermath, I was afraid that the fact that her father drowned would initiate a fear of water in her, but Natasha feels like a fish in water – Alik had time to teach her.)

If they played together, it was impossible to determine who was having more fun. During those ten, fifteen minutes, for which he tore himself away from work, he could entertain her in such a way that, later, when he sat behind his desk, Natasha continued to play by herself for hours. Unlike E.A. and myself, she was permitted to be present in the room when he worked – she did not distract or irritate him.

Natasha remembers numerous episodes of their relations, which slipped away from me, and unquestionably she can describe much better about all of this than I, all that relates to them. The birth of the child, in many ways, bettered Alik. He became softer, more peaceful, and less reserved. Natasha domesticated him.

When she was three years old, we rented a dacha on the shore of the Moscow River. Once she and Alik went off to swim in the river. Having finished my domestic chores, I went to them. On coming to the river, I heard Natasha's loud screaming. Stumbling down the cliff, I saw Natasha's weepy face and next to her Alik frantically running around her, totally at wits end.

"What is it? What happened?"

"I don't know myself. We were playing in the water, and then, as usual, I left Natasha on the shore and went off to swim by myself. As usual, I went into the water, dove in, and then when I swam out,

I found Natasha wailing at the top of her lungs." Our daughter was a very quiet child, so such behavior was absolutely unusual.

"Little one," (that's how Alik always called Natasha), "what is wrong with you?"

"Father disappeared, Father is gone," and Natasha would not calm down.

... A year later Alik drowned...

What was this? An accidental coincidence? Foreboding?

And now before me lay the need to tell Natasha that her father would never return from camping.

"One would not tell a four-year-old child that her father is dead?!" mother asked me this rhetorical question.

"Well, how can I lie? Later when she finds out, she will never trust me again."

It is one thing to make a decision; it is another to carry it through. I racked my brain, what words would I find to prepare Natasha for the news – to let her know that we had become orphaned. Not coming up with anything, I ordered myself, "You will think about this tomorrow. Today's problem is to bring Alik out of Magadan."

Magadan – Seymchan – Magadan

In Magadan, Misha Gelman greeted me, crushed by grief at what had happened. This was Alik's friend from the school bench. After having brilliantly completed his studies at a Moscow institute, there was no placement closer to Moscow than Magadan for this young physicist, Misha Gelman. There, Misha married, settled, and began to live his life. He worked at the Magadan Institute of Geophysics. It was with him that Alik began to correspond concerning finding a partner with whom to go camping. That year, all of the usual camper-friends had somehow run off somewhere and the group was not forged. Alik had been wanting to go to the Far East for a long time. And here a great opportunity opened up – Misha learned about a group of geologists who needed an assistant. With happiness, Alik seized this chance. And now I remember the sorrow Misha experienced, and his feeling of guilt in having organized Alik's participation in this damned geological expedition. I had to calm Misha down by saying that he was not responsible for anything – Alik wanted this himself and was very happy that he ventured out to such a distant place.

The expedition was based not in Magadan, but in the settlement of Seymchan which was located, as Misha stated, "here, not far, some four hundred fifty to five hundred kilometers." Well, by Siberian standards, this is a stone's throw away. One had to travel there on a small local airplane. Fortunately, it departed in some twenty minutes, and Misha was waiting for me with the ticket in hand. One more hour on the road, and finally I arrived at the end point of my journey.

Forewarned by Misha's telegram, the base manager greeted me in Seymchan. I assumed that we would immediately go to Alik, but the manager said that he would first take me to his home so that I could leave my things and wash up after such a long journey. There, he passed me from his hands to the hands of his wife and then left. The wife did not enter into any long conversations, just showed me where I should place my bag and where to wash. When I exited the bathroom, fresh brewed tea awaited me on the table. The manager's wife said, "I know that now you do not want to eat anything, but you must drink some bilberry tea with sugar so as to fortify yourself somehow."

The housewife poured some tea for herself as well, but drank hardly any as she simply sat behind the table and occasionally gazed at me with attentive but not obliging eyes. I asked her whether she knew Alik, and she answered that she had seen him several times and that everybody liked him at the base. She was a woman of few words, and I was grateful that she did not attempt to maintain a mundane conversation for the sake of appearance. When I finished the tea, the wife told me to follow her into the other room where a blanket and pillow, that were prepared for me, lay.

"But I do not want to rest at all. Please take me to my husband."

"You need to gain strength and be in the proper form. This will be the last visit with him since afterwards the coffin will be nailed shut. At least wait until the noise in your ears subsides."

"But I don't have any noise in my ears. There is something like a siren sounding, but this is an external noise. A noise in the ears is not at all like that."

"No, I assure you, it is absolutely silent here. That is a sound in your head."

"How can you know what is in my head?"

"I know," she said calmly but firmly.

She dealt with me as if she were dividing dough – accustomed, aloof, without extra words or gestures. And, like kneaded dough, I obeyed. Before slipping into an uneasy slumber, I thought: "Where does such know-how come from? Do people die like flies here?"

I don't know how long I slept. The wife came into my room at exactly the same time as I opened my eyes.

"Well, here, now you are ready," she said. "My husband will come for you momentarily." Then she added, "Perhaps while he is on his way, you will eat something?"

"No, no, thank you. I do not want anything. And you were correct – now I hear no siren."

"Let's go and drink some tea."

Soon the base manager appeared. Unlike his wife, he was obviously under stress. It turned out that we first had to meet with the Procurator.[28]

"But why do I need to see the Procurator? What dealings do I have with him?" I began to feel impatient, as there seemed to be endless delays.

"Such is the rule of things. Your husband was signed up in the geological expedition as an assistant, and his death falls under the category of a work-related accident. It is imperative to follow through with the formalities."

The Procurator turned out to be a handsome young man of Caucasian descent [29] *wearing a blue uniform. He was concentrated, strapping, and his muscles bulged under his skin. The Procurator said that*

[28]The same as District Attorney in the US. –Editor's note

[29]In Russia "Caucasian" is narrowly applied to people coming from the Caucasus Mountains – Georgians, Armenians, Azeri and a few smaller ethnic groups, such as Chechens. –Editor's note

an investigation had been carried out, which ruled that Alik's death was accidental, and that no one was personally responsible. Another woman sat in the office who was introduced by the Procurator as a court medical examiner. The woman confirmed that there were no traces of wounds or bruises on the body.

"But I don't care what caused this. The only thing that counts is that he is no longer here. Even if you had brought before me someone and said, 'Here is his murderer,' I would feel just the same. It is you, there – judiciary and militia[30] *carry out an investigation and, if necessary, prosecute ... And for me, nothing makes any difference."*

The Procurator's face smoothed out, and the muscle bulges disappeared.

"So, should we close the investigation? You, indeed, will not demand a complete inquiry?" He could not conceal his sigh of relief.

"What for? Is this something that could bring him back to life?"

"Well, in that case, sign right here that you do not have any claims."

And he said it just like that: claims! Some formulaic document was placed before me and, without looking, I signed.

"We will prepare your husband according to all the rules, the base will provide a wooden, as well as a zinc, coffin, and you will have to pay only for transport!" the Procurator declared with an almost happy voice.

Finally, the formalities were over. I was led into a room where the coffin was.

... Alik lay all tanned, looking thinner and younger than when he had left Moscow. The bronze color of his skin set off his grey hair which had grown longer and, due to this, looked more beautiful than ever. The expression on his face showed concentration and looked sterner than before. The softness so typical to people from the intelligentsia *had disappeared as the internal hardness of the character became more pronounced. He had never seemed to me so beautiful as that last time.*

I sat down on a chair which had been kindly put out for me, and

[30] Soviet police. –Translator's note

looked at him for some time – without thoughts or feelings. And then I told him everything. I told him all the things that I did not say during life – how much I loved him, how I had planned to live very long with him, how carefully I had preserved various secrets "for later" when we were old so as to have something to tell him in retirement.

"Why did you leave?" I said to him. "You don't even know at the end what a woman I am. I took so much care to keep our passion unabated, to make sure that our desire didn't fade away into a habit, like brushing your teeth before going to bed. What am I to do with this now, where should I place all of this?..."

I did not even once touch him, as I did not wish to feel the cold body from which his life had exited. My fingers kept his warmth. And I promised to myself that I would raise Natasha in accordance with the principles he had laid down, preserve his manuscripts, and that I would not desert his mother...

After some time, some people came in and said that it was time to close the coffin with zinc sheets. And I, if I so desired, could speak with the geologists who were with Alik in the expedition.

There turned out to be two geologists. They were both men about thirty years old, who were just as tanned as Alik, but alive. They were very nervous, simply shaking. I found out that the group consisted of four people – two geologists and two assistants. One of these assistants was Alik, while the second was a schoolboy who had just finished the ninth grade. In the evenings, while resting, Alik would toss out various problems for him to solve. The teenager discovered for himself mathematics, which he had never seriously considered before. They became closely tied to one another and had planned to continue contact after the expedition and, if everything went according to plan, after finishing school, the boy would apply to Mekh-Mat.

"The kid is in a very depressed state, but if you want, he will come to talk with you," the geologist offered.

"Well, why traumatize the kid even further? And, concerning mathematics, well, this I unfortunately cannot help him with."

A thought passed through my mind: "Here is one more for whom an encounter with Alik was, perhaps, a decisive moment, and his sudden departure from life may turn out to be unfulfilling for the youth."

The geologists told me about the outing. First, in a helicopter, they were dropped off deep in the taiga. Then, traveling lightly, they moved by foot along a planned route. These were absolutely virgin areas. During the entire period (four weeks!), they encountered no human traces, not a single human place of habitation, nor a single living human soul. However, once at sunrise a bear came up to the tent in which they slept. They heard its grunting, and were greatly afraid, despite the fact that they had a rifle, but none of them was a real hunter. Luckily, the bear went away on its own. The journey went unusually smoothly. All the work went according to schedule, as they fulfilled their missions and moved further along the taiga. According to the plan of the expedition, food and equipment was dropped off for them from a helicopter at a number of predetermined points on the map. This was not an easy task, to find the drop offs in the middle of the dense forest, but luck prevailed. Twice during the time of their journey, the helicopter was able to land, so as to collect the mineral samples they had collected. (I knew about this – I had received several short notes from Alik by post.) The last day they were supposed to descend down the Seymchan River and, that very same evening, Alik was planning to fly home. Without much difficulty, they found the rubber inflatable boat, also dropped to them from the helicopter, which they inflated, and began to go down the river.

The river turned out to be very rough, full of rapids, with numerous meanders, covered by tree trunks and branches in many places along its course. Therefore, it was necessary to disembark and to portage the boat by hand. On one of the consecutive turns, the boat's front got caught in the falls and capsized. Three of them jumped to shore and, not seeing Alik, at first did not even understand that tragedy had struck. They began to call for him, but no one answered them. Then they began to look for him, but there was no one near. They found Alik about forty minutes later, on the next bend to which the course of the river took him. But it was too late... Everything had happened very quickly. When the boat capsized, Alik found himself in the water, and he was dragged under the falls. There, under the thick layer of trees, he apparently became disoriented and could not swim out...

In such a way, the eyewitnesses explained to me how he died... They gave me his half-empty backpack and wristwatch. Inside the backpack was a little book written in French by Anatole France,[31] *who was one of his most beloved authors.*

... Alik had a talent of traveling minimally, something that allowed him to often carry the things of other people. In addition, something useful, like, for example, a flashlight or a needle with thread, was never forgotten. There is one very amusing photograph – he and Nikita Vvedenskaya, at an outing, are walking one after another – Nikita's backpack is twice the size of Alik's. Natasha inherited this characteristic from him – when we travel somewhere together, she has half of the things that I do...

Thereafter, Alik and I flew back. We were sent on a cargo plane to Magadan. Some ten people flew with us. In principle, the airplane was not designed for transporting people, but it was equipped with reclined seats which were bolted to the sides of its belly; thus, we sat in two rows facing each other, as one sees parachutists in movies sitting before jumping off the plane. In the middle, at our feet, lay the coffin. The passengers were workers from the geological base. They were flying to the "Large Land"[32] *on vacation and were happy in anticipation of relaxation. This coffin, with the accompanying woman, was rather out of place...*

When, in Magadan, I exited the airplane and walked into the airport, there was Misha, and Victor Palamodov jumped at me with greetings (Palamodych – as Alik came to call him). We barely knew each other. Taken off-guard in the first moment, I didn't even recognize Vitya.

---

[31] Anatloe France (1844 –1924) is a famous French writer. His most celebrated novel, *La Rôtisserie de la Reine Pédauque* (At the Sign of the Reine Pédauque) is a vast tableau of life in eighteenth century France. –Editor's note

[32] Mainland Russia. –Translator's note

"I was sent by the university. Everyone is in absolute shock at what has happened, and I wanted to personally express from myself to you ..."

I looked at him, and my glance undeniably expressed: "Shut up." No sorrowful words, please; otherwise all that will be left of me will be a wet spot!

Palamodych understood and stopped mid-sentence. They told me that the coffin would be reloaded into an airplane headed for Moscow departing in an hour, and, in this way, we would fly further without delay. Misha ran off to straighten out some issues while we remained at the entrance of the boarding area to wait. Having been interrupted mid-word, Palamodych did not dare to open up his mouth. He had sort of a strange appearance – in his beautiful chestnut hair, with noticeable locks of grey streaks, protruded reddish-orange strands. So, as to somehow dissolve the heavy silence, I asked him:

"What is wrong with your head?"

"You know, right now it's summer vacation, and I was playing dress-up with the children. I was called from the university and told of what had happened right at the moment when my daughter was dyeing my hair with henna. I rinsed off my head, grabbed my passport and backpack, and ran off to the university and then to the airport. Only in the airplane did I discover that the dye did not come off," said Vitya, who became incredibly embarrassed that he accidentally touched upon a forbidden topic.

... And I even now still see him in the middle of the Magadan airport – a red-haired clown with a tragic face...

Then Misha came up to us with an airport serviceman. They said that the coffin could not be brought on board the airplane because the baggage compartment was full, and that it would be sent off with the morning flight. Nothing could be done, and we had to be delayed overnight. Misha suggested we go to his place, but I asked to be left alone at the hotel, to which they, with hidden relief, agreed. For people who are not used to death, a person in grief is as an invalid amidst healthy people; they felt uneasy. It is not clear how one should behave – to attempt to falsely cheer up or willingly risk the choice of what can be discussed and what can't be. Of course, it is best of all to drink to oblivion, but is this appropriate to do in front of a newly created widow who, after all, is an absolutely unknown person? There was one more thing – I felt that they wanted to discuss what actually happened, since there were many unknowns in the whole story.

Alik was an experienced and careful rafter – what happened, why, out of the four people, was he the only one who died - someone who sat at the helm, in other words, furthest from the falls? After all, I didn't want to figure out anything in detail. In light of the horror and the irreversibility of what had happened, I did not wish to disturb Alik's soul with a petty, sickening investigation which could not change the tragic end. It was easier for me to think that this was a quirk of fate and not someone's evil will...

Misha and Vitya left for the city, and I remained to stay the night at the hotel. There was no sleep. The strange excitement would not leave me, as I continuously attempted to imagine how it was under the log heap in the water. It seemed to me that with just a little bit more effort, my eyes would see all that Alik went through in the last moment.

Did he understand that he was dying?

What did he wish for?

What was his last thought? ...

Answers to these questions I do not have even today...

And there were still some moot vagaries – partial questions which I tried to banish from my mind – why was the manager of the base so nervous?; why could the Procurator not hide his worry?; why did the geologists simply shake from fear? And this wisdom and experience the manager's wife had about death...

Many, many years since then, I have carefully examined Alik's watch which was given to me along with his backpack. It had stopped on the twelfth. But I was told that the boat capsized on the fourteenth. Why? Alik was very keen with time. This watch, for us rather expensive, he had bought rather recently and was very pleased with it – it worked beautifully. The difference in two days could not be accidental. What happened in actuality? If one were to continue thinking in this direction, his last note to me, which was sent to me from Seymchan, also reads differently. In the first, he wrote that he was very pleased with the expedition, and there were several tender words towards all members of the family. In the second, very short letter, he wrote only about E.A.; he worried very much about her, requested me to take care of her, as if he felt that he would not return and knew that I would not leave Natasha, but his mother – he asked me...

Once, while living in France, I turned on the television, which I normally rarely watch, and found a program about Russia. There was a report about none other than the Seymchan geological station! The program showed an abandoned settlement, which at one time had been a flourishing station, but which had since ceased to exist. What is this? – an accidental coincidence that, unfortunately, is now common and which is to be found throughout the whole country, or a sign sent to me of God's punishment?...

But I do want to believe that it all happened as the geologists related to me, that in the last month of his life, Alik was happy. As I told the people of Seymchan at that time: "If this had to happen, then it's good that it occurred at the end of the expedition and not at its beginning"...

...Alik always had new ideas during his down time; after camping, he jumped to the blank sheet of paper as if he were hungry, as to a morsel of bread.

"What were you thinking about, Alik, as you sat on the helm of the rubber boat; what did you think of, what did you discover?"...

Magadan – Moscow

In the morning, Palamodych and I were the first at the entrance to the plane. We were the first to enter. The passengers following us quickly filled the airplane "up to the knot," as they say. Then we all sat for a long time, waiting for takeoff, but something delayed the departure. Finally, two stewardesses appeared in the aisles. They walked directly to Palamodych and me and, very politely, invited us to follow them into the exit of the airplane and then, just as politely, to descend from the airplane to the ground. There, on the landing strip, awaited two airport service people who explained that the coffin did not fit into the baggage compartment because it had been loaded last, and the baggage of the other passengers was placed in such a way that it did not allow enough space for the casket. When we understood the situation, it was already too late to complain and to be indignant – our airplane was already rolling down the runway. Apologies were extended to us, and we were told that next time, i.e. on the night flight, the loading of the baggage would begin with the coffin and they suggested we spend several hours in the hotel, until the arrival of the plane from Moscow.

I felt slightly faint. Between Moscow and Magadan there are eight time zones, and my biological rhythm had been broken. I could not have said for how many days I had neither slept nor eaten. Palamodych was more composed than I and decisively said that we had to go and eat. There was no restaurant or canteen at the airport, only a buffet, where mainly vodka was sold, but boiled chicken legs were also being offered. Without asking me about anything, Palamodych took two portions. And it was then that I realized that I was very hungry. With greed, I jumped on those drumsticks, and I felt very embarrassed of my animal appetite – well, how could I eat when Alik was dead?! And as if to oneself, I gave an indulgence – it was necessary to get nourishment since, otherwise, there would not be enough strength to get to Moscow. "But, in memory of that weakness, I will never again put chicken in my mouth!" I quieted my conscience.

... However it may be – now I do eat chicken once in a while... Everything passes... well, but not without a trace...

Later, at the hotel, we slept like the dead; we would get up, go to the buffet, eat those same chicken legs, and then go back to sleep. Several times, the administrator knocked at the door and asked whether we were ready to vacate the room, since passengers from the Moscow flight had arrived, and the hotel was fully booked. In the end, we left the room to the happiness of its next occupants who were standing outside the door.

When, at the airport, we gave our ticket for registration, it turned out that we had long been expected, and immediately, before the announcement for boarding, we were led to the airfield. There, next to the airplane, stood a transport vehicle with Alik's coffin. The embarrassed airport attendants said that they especially brought us here to show that they had attempted to do all that they could in their power – the loading of the baggage on this flight, indeed, began with the coffin, but its dimensions did not fit the standard, and it did not fit into the airplane – it was impossible to bring in and turn around the coffin, so as to place it alongside the area of the baggage compartment, since it turned out to be too long; for this same reason, it could not be left in the perpendicular position since then the doors of the plane would not close... Yes, the people of Seymchan were not stingy with material for Alik, the large zinc bed was made for him! ... We were assured that this could be rectified on the following morning when the workshops would be open, and then the zinc cover will be shortened and made smaller on the sides.

It was senseless to return to the hotel, since we knew that there were no vacant rooms, so we went off on the transport vehicle, along with the coffin, to the warehouse.

The warehouse was, in actuality, a fenced-in area on the airfield, inside of which was wooden flooring covered by a canopy approximately 200 $\times$ 15 meters, occupied by cargo that had arrived by plane, and some twenty meters away from this site stood the night watchman's stand.

There were three night watchmen. They played cards and gnawed on apples. They offered some apples to us, saying that they would rot anyway at the warehouse. But we were not permitted to stay at the watchmen's stand, because this was prohibited by the rules, as it was explained to us. In reality it was clear what the situation was – why the hell would they need in their warm company these two funeral party members. Oh, well, it was possible to understand them. They had already expressed extreme kindness as they found a place under the canopy for the coffin. We went there ourselves and settled down near Alik. It was dark, quiet, and damp. A warm drizzle was coming down. We sat amidst cardboard boxes with refrigerators, televisions, and some other appliances. Almost all of the equipment was placed under the canopy – only a few boxes here and there stood in the warehouse yard under an open sky. There, also, stood uncovered wooden crates of apples. These apples had begun to rot under the rain, and a sweet, suffocating smell emanated from them. I imagined how these apples had been brought on a train from Moldavia or the Crimea to Khabarovsk and then transported by plane here. And now they were rotting.

"Well, what kind of economy can withstand such a thing?" I thought. "And what pain this scene would bring to Alik's soul."

* * *

I do not know whether Alik could foresee that the USSR would collapse during our generation, but the fact that the economy was taking its last breath, that the country was irreversibly moving into a dead end, he knew very well. He perceived what was happening as a personal deadly disease. A Jew according to his passport, he felt that Russia was his only motherland and was tied to it with all the passion of his soul. With unspeakable pain, he spoke of how predatorily the forests were being chopped down and the rivers silted in, how nature cannot survive this debacle brought on it by the Soviet bureaucrats who do not value anything and feel no responsibility for what they do.

Once he came home in a state of significant excitement and distress. "I received an offer to leave the university and fully switch to ecology. They offer a pile of money and paid trips wherever and for however long I want. So what do you think?"

"A pile of money would be very useful for us and, with a formal obligation to travel throughout the country, you would feel morally at ease to escape from your dear mommy. But all of this governmental concern about ecology is totally superficial. The last say will always belong to someone 'from the top.' At this time, you only confront the university administration, and, even so, how many nerves you waste at this. At least you sit behind your writing desk and feel like a free person and do not depend on anybody. Working in ecology, you would be tied by your arms and legs, you will be constantly struggling, worrying, but still unable to do anything."

As always, silence was an answer to me. Alik walked off to think in the other room... He did not leave the university...

* * *

It was not only the misfortunes of Russia that Alik perceived as his personal sorrow. It was as if he held his hand on the pulse of the planet, and each of its disasters he felt as a blow made against him personally. I will describe one of the many instances which I had witnessed.

Once he and I were strolling through the forest and were, as I felt, in a wonderful state of spirits. I was taking in the scenery of the various flowers, foliage, and butterflies all around. With a sudden glance at Alik, who was as usual deep in thought, I suddenly discovered that he had a very anguished face.

"What happened? What is wrong?"

"Well, how can I be happy, and even simply live in peace, when at this very time, Pol Pot is destroying millions of people.[33] The world 'did not notice' when there was a genocide of the Armenians in 1915,[34] then the destruction of the Jews during WWII, and now, the same thing is going on in Cambodia, but the world community, as before, only stands idly and watches..."

... Everything touched him, he filtered everything through himself...

* * *

When people of our circle began to leave the country, speaking of himself, Alik said, "I will go in the last wagon of the last train, but only if I am deported. Instead of America or Israel, I choose Birobidzhan."[35]

---

[33] Pol Pot was the ruler of the *Khmer Rouge* and the Prime Minister of Cambodia from 1976 to 1979. During his time in power Pol Pot instigated a policy of mass extermination of intellectuals and other "bourgeois enemies" in an attempt to purify the Cambodian people as a step toward a communist future. –Editor's note

[34] Forced mass evacuation and related deaths of million plus Armenians during the government of the Young Turks from 1915 to 1917 in the Ottoman Empire. –Editor's note

[35] Birobidzhan is the administrative center of Jewish Autonomous Province in Soviet Far East. This Province was established in 1934 as a culmination of Stalin's "theatrical" project to provide a home for Jews in the Soviet Union. No mass Jewish migration ever took place. At the maximum about 40000 Jews lived there. In late 1936 the project came to a crashing halt with the onset of Stalin's mass purges. The entire region was declared out of bounds for normal citizens and NKVD (the Soviet Secret Police) was given control of it. Currently it is an underdeveloped area of Russia with a tiny Jewish minority in a mostly Russian population. –Editor's note

I believed that emigration was a must and quietly pushed him. I should mention that Soviet power greatly helped me in this – the situation in the country (and, in addition, at the university) became more and more unbearable. And, furthermore, I had a strong argument – I insisted that we must do it for Natasha. Alik would sigh, become silent ... but in fact he did make some sort of steps. He was offered a permanent position in Poland and, after long discussions, we decided that this could be an intermediary variant – to depart for a prolonged period of time to the country of the Warsaw Pact and there, depending on the circumstances, go further west or return to the Soviet Union. The question that arose, as usual, concerned his mother, with whom we would have to live if we left the country. I took upon myself the responsibility of getting along with her peacefully, of which Alik was highly skeptical. He asked me not to disturb him and give him the opportunity to peacefully finish working on his book *Introduction to Superanalysis* which he planned to finish in the spring of 1981... He did not have time...

* * *

Alik could easily have registered himself as a Russian. When, in his sixteenth year, he came to fill out the paperwork to receive his passport, he was summoned by the head of the department in the militia (police) and asked:

"Your father was Russian, so why do you write in your application that you choose a Jewish nationality?"

"I was brought up by a Jewish mother."

"But do you understand that by choosing to be a Jew you just ask for more complications in your life? And you have the legal right to avoid all this!" The representative of the powers-that-be would not relent. (It was neither out of the kindness of his heart nor because of his particular disposition towards Alik that he engendered this discussion – at that time, this was the state policy – to limit the official numbers of Jews in the USSR by all means possible.)

"Jews live somehow in this country, and so will I," Alik answered.

... All of his life he paid for this decision, and he never regretted his choice...

... If someone would ask me to list the key traits of his character, I would have said nobility and manliness...

* * *

Once I asked Alik:

"Do you forgive Valera Nikolsky's drunkenness for his kindness?"

I was shocked by his response. "Valera drinks to forget, because he is up to his neck in the dissident movement."

"What does 'up to his neck' mean? You have always participated in this movement."

"I? No, never," Alik expressed himself in an unusually lively manner. "I only once signed a letter in defense of my friend and colleague (he meant the letter in defense of Esenin-Volpin [36])."

"And why? You hate the current regime."

"Because I do not find a positive message amongst the dissidents. All the fighters for the citizens' happiness brought those same citizens only sorrow and rivers of blood. There is no major difference between Pugachev [37] and Ulianov (Lenin) – all of them are criminals."

"And what about the Decembrists?"[38]

"Well, these are saintly idealists. They were lucky, since the uprising was extinguished at its root, and they had no time to ruin anything."

"So this means that with their innocence they were obliged to the tsar?"

"Yes. If Nicolas I was not firm, the Decembrists would not have been sainted."

---

[36]Alexander Esenin-Volpin was a prominent Soviet mathematician, a notable dissident, political prisoner and poet. A pioneer of the human rights movement in the USSR, Esenin-Volpin spent fourteen years in prisons and exile. –Editor's note

[37]Emelian Pugachev (1742-75), a Russian peasant leader, head of the peasant and Cossack rebellion against Catherine II of 1773-74.

[38]The (failed) Decembrist revolt was attempted in Imperial Russia by army officers who led about 3000 Russian soldiers on December 14, 1825. The officers were particularly incensed that Alexander I had granted Poland a constitution while Russia remained an absolute monarchy.

"In other words, there are no exceptions – 'Russian rebellion is senseless and merciless' always?"

"Yes, and Pushkin understood this beautifully. It was not only because of kindhearted friendship that Pushchin[39] visited him in Mikhailovskoye.[40] He went there at the request of the Secret Society, so as to report the preparation for the uprising, but Pushkin refused to participate in it. And Nicolas I knew about this, and, for this reason, he got closer and kinder to him. It is known that during the investigation of the Decembrist revolt, he called on Pushkin, to see him, and they had a long dialogue. We do not know what Pushkin told the tsar, how he imagined the creation of a decent way of living for the people – and for all of us this is a big loss."

"Well, you are smart – you invent a positive program."

"I have thought about it for a long time, but nothing comes out of it. In any given situation, the people always suffer."

"And, if you could propose a positive program, would you go into politics?"

"Absolutely. Without any hesitation."

"So this means that you run off into mathematics as into drunkenness?"

A bit hesitant, Alik answered, "To some degree – yes."

* * *

The night counted out the hours of its reign at the warehouse yard. Palamodych and I sat there quietly near the coffin.

"Do you know," I broke the silence, "All of this is not an accident that we just did not fly out of here. Just think, we were taken off the plane three times! And also in Domodedovo, they tried to stop me from going. It is simply that Alik does not want to leave from here. What is he to do there, in that stuffy Moscow from where he always attempted to escape, where he was not appreciated, and which bothered him in various ways of life and work. He is a scientist

[39]A member of a Secret Society which was behind the Decembrist revolt. The first such society, "The Union of Salvation," was established in St. Petersburg in 1816. –Editor's note

[40]The Pushkin family estate. –Editor's note

of world renown, someone who taught numerous graduate students and PhD's, someone who created his own school, discovering a new direction in science, formally was not even a professor, and so left life as a senior scientific fellow! And here now, the university, bereaved by his death, has sent you and allocated money. Should I be thankful for this?! I just want to send them all to hell! And I warn you that I will not permit any officials at the funeral – I will simply put them out. You'd better warn them so that they do not dare to come. Alik lived amidst them as if in prison and, based on the irony of fate, here, in the land of the gulag, he felt free. Can you imagine," I continued, "how wonderful it would be to bury him here, on the high bank of the sea, erect a little pyramid, the type that are placed on the tombs of the unknown soldiers, because he, in reality, fought all his life – quietly, without a show, but no less selflessly and courageously. At the top of the pyramid, I would place not a five-cornered star, not a cross, and not a Star of David, but something like a light bulb; and even the pyramid itself should be covered with phosphorous paint so that it would glow like a lighthouse, to be seen by the ships going to Magadan..."

"And will you do this? Will you have the nerve?" asked Palamodych.

"No, I will not have the nerve," I honestly confessed. "I will not be able to withstand the pressure of public opinion, my gut is too thin. And to excuse my own weakness, I have another reason – I have to bring Alik to his mother so that she can say farewell to her son..."

And we once again fell into long silence...

And the rain continued to drizzle – it was humid and stuffy. It seemed that every drop, like a small hammer, pierced the air to the ground. And together with the thickening air, the stench of rot continued to grow. The apples decayed with amazing rapidity. And we, at this Aeroflot warehouse, were participants in the Divine Comedy, that same part when "everything" turns into "nothingness"... From that time, I know the smell of death – death smells of rotting apples... The sweet smell brought on a high, everything in my mind became confused. Perhaps such a sensation is experienced by hookah smokers... I stood up from the wooden pavement on which we sat, saying, "I'm going to walk around a bit."

Between the warehouse and the actual runway was an area overgrown with some sort of low-growing vegetation. I went along this "neutral lane," tripped, and fell. It did not hurt, but it was very demoralizing. And it was then that I broke down. I began to cry aloud. I rolled around on the ground, shook in hysterics, cried, and wailed. Then I passed out. I do not know for how long this continued. On regaining consciousness, I realized that I lay on my stomach, with my face buried in the ground. My nose was tickled by some unknown smell. It was a tart, sharp, invigorating, and fresh smell – I wanted to breathe it in endlessly. All around was pitch darkness, but, for some reason, I knew that this smell was like a green chlorophyll. So here it was, that fragrance of the taiga *for which those young tourists came! Alik also ran off for the same smell! For me, it became the scent of life – I greedily breathed it into myself. And, as if on a screen of a supermodern medical machine, I saw how it cleansed my mind, how the blanket which covered it faded away. Now the brain was fully awakened, pulsing from the circulation of fresh blood, as I continued to breathe in this enlivening smell, and I did not wish to stop. I stored some for the future, for the rest of my life...*

Finally, I arose, ripped open my chemise, and wiped off my face, on which there was a mix of tears, snot, and saliva, somehow automatically noting that I can wipe myself off so nicely because the shirt was made out of pure cotton and that Alik was right once again when he taught me not to wear synthetic fabric – what would I do now wearing a nylon lacey chemise?! – and I walked towards the warehouse.

"Where were you for so long?" Palamodych sounded worried.

"I was learning how to live further," I answered and then said, "You and I are wasting our time, since when the workshops open, while they work on making the coffin smaller, we will be late for the morning flight and lose another day."

"So what should we do? I could have fixed everything myself, since I can do anything with my own hands, and it is for this reason that the university sent me here; but, in these unforeseen circum-

stances, I don't have any tools to work with."

"What this means then is that we need to get them."

I went off to the night watchmen. They said that they did not have any tools, but perhaps the firemen could help me. Their station was located some 300 meters from the warehouse. I went there. I had to knock for a long time – the firemen "slept like firemen," but finally, having awoken and heard my request, they immediately gave me various hammers, a sledgehammer, chisels, and God knows what else. They sincerely regretted that they could not help me themselves because, during the time of their shift, they had no right to take even one step outside of their station. With all these tools, I returned to the canopy area, and we began to work. Palamodych began to pound on the zinc coffin with a sledgehammer, so as to diminish its volume by reducing the space between it and the wooden coffin. Under the blows, the zinc sheets bent through, and the nails that held them together began to give way, which compromised the integrity of the coffin. It became necessary to "sew back" the zinc sheets. Neither we nor the firemen had anticipated that, aside from the tools, we would also need nails. We did not wish to disturb them again, so I began to straighten out the nails as best I could on the rock that was located inside the yard. Palamodych, indeed, turned out to be a master handyman as he was able to fasten together the coffin by closing the cracks with these bent nails. Towards morning, the palms of his hand were all calloused, but the coffin corresponded with the standards for transport. There was nothing left to keep us from leaving Magadan. We departed with the morning flight.

The End and Beginning

Three to four young mathematicians awaited us in Domodedovo to accompany the hearse. Perhaps I should have known them but, at that moment, I did not recognize anybody. Alik was taken to the morgue at the Sklifosofsky Institute.

It was July of 1980. Moscow was preparing for the Olympic Games. All around were militia posts, and Muscovites were advised to leave the city if they could; entrance into the capital was strictly limited, and there were very few cars on the deserted streets since the traffic was restricted. Our vehicle was often stopped and our documents checked.

... Yes, Alik was an inconvenient person during life, so he remained inconvenient in death...

Much later I found out that a special "Berezin Funeral Committee" had been organized, and, in hindsight, I understand the love, care, and attention that was expressed towards him. All those whom I did not notice and did not thank – I bow down low with gratitude... The following day was the funeral. Much time had passed since Alik's death, and the farewell could be done only in front of a closed casket. But E.A. said that she wanted to see her son and requested that the coffin be opened up for her. She and I came into the morgue. Blue-lily spots were already forming on Alik's face. E.A. came up to the coffin and began to stroke this changed face. The skin under her fingers began to peel off. She moved her hand to other places, and there again pieces of the face began to fall off.

"God, isn't she afraid?" whispered the worker of the morgue who stood near me.

"She is not afraid," I answered in the same whisper. "She is a pathologist-anatomist."

Then the coffin was closed and the farewell began. It was warm, sincere, and inexpressibly bitter. Only people very close to me and Alik were present. First to speak was Volodya Tikhomirov, and he said all which I, as a friend of the family but not a mathematician, could not say – that Alik was on the rise, ready for takeoff, that his years had given him wisdom which did not blunt his creative potential, that, among his contemporaries, he had retained his youth better than others and remained youthful in his work, and that the scientific community had only begun to know the total worth of his work, and that he was on the threshold of fame... Then spoke Nikita Vvedenskaya, Victor Palamodov, and someone else, but, diving into my own thoughts, I did not hear anybody else...

...Did not live long enough, did not love long enough, did not see his daughter grow up, did not have time to do everything which he was capable of, did not wait to receive the recognition he deserved...

And then it was necessary to continue my life. I moved to Vinnitskaya Street since it was impossible to leave E.A. alone. We did not get along during Alik's life, and now we had to live together when he was no longer there.

We attempted to bring order to his papers, to figure out the manuscripts. Friends, mathematicians, and Alik's students came over, and the question of publishing his unfinished works was brought up.

... Soon after Alik's death, thanks to the concern and effort of Kirillov, Palamodov, Shubin, Schwarz, Ogievetsky, Fradkin, and Leites, these were published: Introduction to Superanalysis, Introduction to Algebra and Analysis with Anticommuting Variables, Shrödinger Equation... *Having established what needed to be done in Alik's memory, the mathematicians left to finish their summer vacation.*

And E.A. and I remained together.

I did not wish to see anybody; the television static irritated me; and we didn't even turn on the radio. We laid out Alik's still unclaimed papers, as best we could, and then cleaned the apartment. It became light and empty. I had the impression that even dust did not accumulate. It seemed to me that we lived inside a glass prism with thick transparent walls which did not let in any images of earthly life. But they were unnecessary anyway. It was as if we had relocated to Purgatory, but did not notice the shift into another world...

In her "previous" life, E.A. had spent most of her time reading. But now she could not read – she was unable to concentrate on anything except for her sorrow. I understood that she could not be left alone with her own thoughts, so I offered to read aloud. E.A. loved Chekhov. His reserve, intellectualism, hidden humor, and the deep layers of contradictory feelings behind banal words were close to her heart. We ended up on Chekhov's short stories. E.A. especially loved "Ionych" – we read it several times. And we both felt that Mandelshtam[41] *resonated with our mood:*

[41] Osip Mandelshtam (1891 – 1938), one of the most cherished poets of the Russian *intelligentsia*. He wrote verses distinguished by classical restraint, majestic conciseness, and sonority. His late poetry draw a gloomy picture of the Soviet reality of the 1930s. Arrested in 1934, he was exiled first to the Ural region and then to Voronezh. In 1938 he

What to do with the wounded bird?
Solid, restrained, the silence:
the bells, out of the misted
bell-tower, have been stolen.

And the heights stand,
like a white empty turret,
mute and orphaned,
of mistiness and quiet.

And then we would go to the kitchen to drink evening tea, loaded up with tranquilizers prescribed for us both, before going off to our rooms. At that time, I did not know that totally different medicine should have been given to E.A.

... Soviet psychiatry was marred by its dealings with dissidents, who were put into psychiatric wards if, for various reasons, it seemed impossible to send them to prison. In addition to that, there was a less noticeable, but quite harmful, influence of official state policy on this branch of medicine. In agreement with official morality, the "builder of communism," by definition, could not be anything but an optimist, and the concept of clinical depression was practically wiped out from medical vocabulary. Only in France did I find out that depression amongst the elderly leads to senility and memory loss. E.A. needed to be treated not with tranquilizers but with antidepressants...

At the end of the summer, I went to Riga for several days to bring back Natasha. During my absence, E.A. left the house and was unable to find her way home. She forgot her address and last name, and remembered only that she had a son named Alik and requested that he be called. On my return to Moscow, I managed to find her in the Kashchenko Hospital.[42]

... E.A. often recited: "God protect me from going mad, I'd rather go a-begging on the tramp."

---

was rearrested, sent to a *gulag* camp, and died in harrowing circumstances in Vladivostok. The following verse is from Mandelshtam's first book, *Kamen* (1911), translated by A. S. Kline. –Editor's note

[42]The largest psychiatric hospital in Moscow. –Editor's note

Her prayer was not heard... She outlived Alik by three and a half years...

As we know, unusual encounters occur in life. I was destined to find out that no less strange coincidences occur in death. Some two weeks after Alik's death, Sveta called me and said:

"Why are you sitting around there like a nobody? You cannot go on that way. Life continues."

"Well, and how does life continue?"

"Yes, here, now Vysotsky[43] *died," said Sveta as she herself became uneasy by her own voicing of this absurdity... Uncontrolled laughter came over me. And this news, as a counter-shock, brought me back to real life.*

Several months later, in connection with formalities concerning the publication of Introduction to Superanalysis, I had to get a written statement of Alik's will.

... At that time, it was considered that a Soviet citizen could not have any private (at that time they said "personal") property. For this reason, the only office which handled issues connected to the inheritance of citizens in Moscow and in the Moscow oblast' (region) was Notary Office Number 1, or, more specifically, one of its four branches. The office was located on Kirov Street in a building erected in the early twentieth century. The inheritance branch occupied the top floor, the fourth. Considering that the ceilings in this building were some five meters high, and there was no elevator, the "ascent" was equivalent to at least seven floors in an ordinary building. To climb up there was a hardship in itself. But this was a lesser one. It turned out that this floor, in contrast to the rest of them below, was crammed with people. In the capital and its surrounding region, there were still enough people who had some sort of possessions or intellectual property that their inheritors could fill up a spacious

[43]Vladimir Vysotsky (1938 – 1980) was a great Russian *bard* – song-writer, poet, and actor, whose work had an immense impact on Russian culture in the 1960s and 70s. He continued and developed the tradition of Bulat Okudzhava and was cherished in *intelligentsia* and youth. His songs were perceived as a breath of fresh air in the stifling atmosphere in which Soviet people were forced to exist. His work was largely suppressed by the official Soviet establishment. –Editor's note

corridor of the old building to the point that there was no room even for an apple to drop...

My visits to the notary followed one and the same routine. After work, as a rule overburdened with heavy bags of groceries, I would drag myself to this fourth floor, practically the seventh, and would sign up for an appointment. Thereafter, on the appointed date, I would take a day off and would spend hours waiting in line; afterwards, the notary, in just two minutes, under idiotic pretexts, would kick me away like a football.

In the end, I staged a scene. At the height of our conflict, the notary grabbed me by the shoulder and ordered me to go with her somewhere into the depths of the corridor, to a door with the label "Employees Only." We entered a room which was approximately sixteen square meters.[44] *Two-thirds of it, from the floor to the ceiling, was occupied by a mound of files dumped into a heap, each containing an individual's inheritance documents. The furniture in the room consisted of a single Vienna chair with a broken back.*

"Here" – the guardian of the law told me – "amidst these files is yours. We don't have an archivist. If you wish, find it for yourself."

She left, and I began to sort through and pick over the files. In this way, I spent four hours bending down and standing up. Suddenly a file slipped down from the mound, and I almost stepped on it! On the file there was a label: "Vladimir Semenovich Vysotsky." Carefully I picked it up and placed it on the Vienna chair. Finally, I found Alik's file. Then I sat down on this half-broken chair, placed both of the files on my knees, and began to quietly rock just as one does when lulling a baby to sleep. I sat there and thought of both of them, two such different people. One was a poet-singer cherished by the people of a huge country, while the other was a scientist known throughout the world – both half-blood Jews, they were loyal to their homeland, Russia, until their last breath, suffering their misfortunes together with her, and living their short lives under the weight of a state power which treated them like a stepmother treats her unwanted stepsons...

[44] Approximately 160 square feet. –Editor's note

... And all around lay other files, with other fates, and before me lay a materialized postulate: "everyone is equal in death" ...

Later, I stood up, left Volodya Vysotsky on the chair, and went to fill out the inheritance paperwork...

And there was one more "virtual" encounter with Vysotsky, when I buried the urn.

According to family tradition, Alik was cremated. We had a gravesite at the Vostriakovskoe Cemetery. But it seemed to me that it would be more fitting if Alik were laid to rest alone, away from everybody, something that he deserved at least after death. Officially new burials were prohibited at the Vostriakovskoe Cemetery, but the university helped once more. It was once again Palamodych who helped me – he brought me the appropriate paperwork, typed on official stationary, with the seal and signature of the Rector. After the inevitable hassle and runaround between the administration of the cemetery and Moscow City Hall, permission was granted.

I picked out a place for the grave at the intersection of two paths of secondary importance. It appeared to me that it best suited Alik – alone but not lonely, since people walk there from time to time. One can hear their voices, so he could continue to participate in life.

I did not wish to erect a metal fencing along the perimeter of the site. Alik was behind a grate all his life, so let him lie free in his death. And the grave itself appeared to me like a replica of Russian nature, which Alik loved tenderly, amidst which he strolled... The entire area of the lot was made into a high mound which was covered with green undergrowth. A birch and a willow that had just begun to sprout were kept intact. For the tombstone I wished to lay a simple, unworked boulder, like those encountered in the central regions of the Russian plain, but it turned out that it was extremely difficult to get one. Someone said that for Vladimir Vysotsky's grave, geologists brought a meteorite from the taiga *on their own backs. Marina Vladi,[45] Vysotsky's wife, wanted to have it placed on his grave, but the poet's parents preferred something mindless, with a guitar and horses standing on their hind legs. "Ask Marina to give you the*

[45] A well-known French actress. –Editor's note

meteorite," my friends suggested. At first I became intrigued, but later thought: "This meteorite is soaked through with the geologists' love for Vysotsky. Alik deserves his own personal stone, despite the fact that it may not be a visitor from the sky."

In the end I came to an agreement with the workers of the cemetery, who found the type of boulder I was searching for. They placed it at the right side of the mound, at an angle away from the trees. In such a state, the grave stood more than ten years, but the earthen mound had to be mended many times, as it settled. The green grass required constant attention. It became apparent that without continuous care, the grave would be quickly ruined. But I am not eternal, and as for Natasha, well, she is far away, living in another country.

Having thought about it, I decided to cover the mound with marble. The all-knowing workers of the cemetery said that the grave of an adult person should be covered with black marble. But Alik had the light soul of a child, and I preferred white marble. So it turned out to be a light plinth, some half meter high, with skewed asymmetric sides but in harmony with the surrounding landscape. The plinth seems to rise up out of the ground naturally. The boulder was situated in its previous place, to the right of the plinth. Its entire surface was laid out in two lines, made of the same white marble, but broken up into small, irregularly-sized pieces – as if it was an abstract image of two prostrate birches on the grave of Berezin.

To the left side of the white plinth, as if breaking out of the ground and piercing through the marble facing, rising high into the sky, are two living, straight trees. At their base, there is the following epitaph:

BEREZIN
Felix Aleksandrovich
Mathematician
1931–1980

Well, this is it... the earthly journey of Alik has ended...

* * *

Once in the cemetery a few women passed by me. One of them, turning to her companions, said, "Now we are going to Berezin..."

Unwittingly I squeezed into their conversation, "I'm also going to see Berezin. Were you somehow related?"

Embarrassed, the woman answered, "No, we don't know him, this is just our reference point..."

And I thought, "This is Alik's destiny – to give people direction." Coming to his graveside I do not wish him to rest in peace; I try to soothe him, telling him, "Now you have more followers! Recognition has come to you! People understand and appreciate you!

La Rochelle
July 2005

Glimpses of Felix Berezin's Life

Alik Berezin as a schoolboy, 1940s.

Early 1960s

Hiking

On a hiking expedition

Berezin engrossed in his beloved mathematics

Esfir Rabinovich, Felix Berezin's mother, 1970s

Late 1960s

Elena Karpel, late 1970s

Alik and Natasha

Late 1970s

Boris Voronov, Dmitri Gitman and Igor Tyutin (right to left) pay homage at Felix Berezin's grave in 2005.

REMEMBERING ALIK BEREZIN *

V. P. MASLOV
Moscow State Institute of Electronics and Mathematics
Department of Applied Mathematics
RU - 109028 Moscow
Russia
pm@miem.edu.ru

When I was a third-year student at the Physics Department of Moscow University, I made Alik [a] Berezin's acquaintance through my close friend N. Korst, who lived in the same house with him and had known him since childhood. Alik's nickname among the boys in the courtyard was "Beribes" (beri, bes in Russian means "take it, devil" and sounds like his last name). He lived in a two-room apartment with his mother, who adored him and was very supportive of his passion for physics and mathematics. She was a good cook and enjoyed playing the role of hostess. She was a regular listener of Radio Israel and showed keen interest and emotional involvement in what was going on there.

Alik used to study with his window flung wide open both in summer and winter alike, the table moved up close to the windowsill. When he was thinking about a problem, he used to pace back and forth in his room. He would often think while strolling in the woods. Once, when he was visiting me at my dacha, we took a walk in the nearby forest, talking as we went along the paths. After a short pause I asked him about something. He answered: "You are interfering with my thinking. Don't disturb me for a while, I am working,"

---

* Translated from Russian by A. Sossinsky, Independent University of Moscow, Bolshoy Vlasyevskiy Pereulok 11, 119002 Moscow, Russia; e-mail: lifrc@mccme.ru. Reprinted from *American Mathematical Society Translations*, Series 2, vol. 175, pp. 225-226, 1996, with the kind permission of the Publisher.

[a]Felix Aleksandrovich Berezin was known to his friends as Alik. –Translator's note

(on some problem in mathematics, no doubt).

We often talked about politics and politicians. I remember that he surprised me by saying that Khrushchev was the only guarantee of our security. That was all the more surprising since both his mother and I had a feeling of disgust and revulsion towards Khrushchev. Alik seemed to have strong misgivings about the future.

Many years later I gave him my first book as a present. He read the book and later returned it with a dedication of his own, which was a funny parody on my own dedication, which I had thought rather touching. The returned book also contained numerous reasonable remarks and some witty ones. One passage in the book was commented on in the following facetious way: "This is a piece of voluntarism, which is censured by our beloved Party!"

We had lengthy discussions on mathematical craft, on the major trends in mathematics, and about the leading mathematicians. He was of the opinion that one should take up two widely differing branches of mathematics and do research in both with the utmost vigor and professionalism. According to him, this would benefit both branches.

He "awarded" military ranks to our "great" mathematicians. Thus, in those days he "promoted" L. Pontryagin to the rank of colonel, and I. Gelfand and M. Krein to that of general. The reason was that, as he explained, unlike I. Gelfand and M. Krein, Pontryagin had no army (of disciples and followers) of his own.

In general, he divided mathematicians into "essentially creators" and "essentially teachers." His appraisal of himself was rather modest. He used to say that his own destiny was to teach and not to create. Of course this was not true, but it indicates how devout he was in his calling as a teacher. Moreover, in his heart of hearts he realized that his research work was very important. He was much aggrieved that some of his work was not immediately accepted for publication. He felt that it was a shame not to be able to put his ideas across to the reading public, since he realized that he understood the overall picture better than his more famous competitors.

Generally, his attitude had always been absolutely independent and ahead of its time. Even in 1979, a year before his tragic death, he was irritated at the mathematics community for not being able to grasp where the frontiers of science were. In a way, he resembled Nabokov's Pnin; that is to say, he always looked somehow unprotected. He was not a fighter. People who live before their time die early.

Moscow
1996

ON BEREZIN *

M. A. SHUBIN

Department of Mathematics, Northeastern University
360 Huntington Avenue
Boston, MA 02115, USA
shubin@neu.edu

These are personal notes concerning my acquaintance and relationship with F. A. Berezin.

I believe that the fact that I met him and was able to talk with him and learn from him for a considerable period of my life was one of the best gifts of fortune to me.

We first met in 1967 during a Summer School on Representation Theory and Spectral Theory in a small summer resort called Zagulba on the Caspian sea near Baku. I was then 22 and had just completed the first year of my Ph.D. studies (my advisor was Professor M.I. Vishik). I guess that there were not very many single rooms in the place where we lived and certainly I did not expect the privilege of being given one. So there were three participants of the school in my room, the two others being F. A. Berezin and another graduate student of my age, Grisha Litvinov. Berezin was then 37, and from our (Grisha's and mine) point of view, very famous and deserving the highest respect. But for the administration of the school, he was still not up to the level of those who rate a separate single room.

Anyway all three of us lived peacefully in this room and, as far as I understand, Berezin first attentively observed the behavior of his

* Translated from Russian by A. Sossinsky, Independent University of Moscow, Bolshoy Vlasyevskiy Pereulok 11, 119002 Moscow, Russia; e-mail: lifrc@mccme.ru. Reprinted from *American Mathematical Society Translations*, Series 2, vol. 175, pp. 226-228, 1996, with the kind permission of the Publisher.

roommates and our occasional mathematical conversations. At some point he asked us the following question:

"Do you want to learn quantum mechanics?"

"Of course we do," we answered. What followed was totally unexpected for Grisha and me.

"Let us write a book about it then," said Berezin. I do not know precisely what Grisha felt about this, but for me this proposal was astonishing. Authors of all books seemed persons on the level of demigods, not less. To see my name on the cover page of a book seemed something totally out of reach. But of course it was an attractive proposal. I could not believe that Berezin was serious, inviting such inexperienced people to participate in an act of creation. But he definitely was serious. He explained to us what he wanted us to do. We were expected to transform into a book some existing lecture notes of his and add something that he would explain to us. This is how my relationship with Berezin started.

Unfortunately, progress on the book was very slow. I was virtually illiterate in most aspects of physics, spectral theory and all the other things that were necessary. The lecture notes contained almost no proofs, so that we had to invent them (this was much easier than to find them in books and papers by other authors because of the poor situation with the libraries, not to mention the complete absence of copying facilities). Berezin was ready to help when we asked him, but his proofs usually contained new gaps to be filled and we did not dare to ask him for too many clarifying explanations. Grisha quit after learning the basics of quantum mechanics and writing a chapter about them. I continued, and as I see now this was one of the most profitable learning enterprises in my life. I acquired an active knowledge not only of quantum mechanics, but also of functional analysis, spectral theory, and representation theory. Moreover, my fear of physics and physicists gradually dispersed. I was reluctant to come to the Berezin seminar, where many talks in physics took place, because I was afraid that I would not understand anything. But once, before a physics talk, I shared my reluctance with Berezin and he said:

"Do you know already that a physics quantity is a self-adjoint operator?"

"Yes, I do," I answered.

"Then you don't need anything else," Berezin assured me. Listening to this talk I found that this was almost true. This situation was repeated again, and after a while I was not scared by physics talks anymore, realizing that actually not so many things are needed to understand them.

Many times I was a guest in Berezin's apartment, to talk about mathematics (there were no offices in Moscow State University, so people worked mostly at home). He also often shared his nonmathematical views about different aspects of life with me. He was a man of courage, never stepping back from his moral principles, although he was never an overt dissident. Sometimes we went skiing together. But this was too difficult for me, because he was much stronger and his level of fitness was incomparably better than mine.

Unfortunately the book was never finished in the form which Berezin had planned. The reason was that I was lazy, and besides I felt that I could do good things which might be more interesting. A preliminary version of the book was published in Russian in 1973 as "Lectures in Quantum Mechanics." But this was a very small part of what was planned. Then the work slowly advanced. Amazingly, Berezin never became angry at me about it (although I definitely deserved severe reproach). Only after his tragic death did I realize how criminally lighthearted I had been about the book. It became suddenly clear to me that I had lost a unique opportunity to learn a lot more from Berezin (being sure that I still had plenty of time) and that the book would never be completed at the desired level of quality. Then at last, strongly urged by A.S. Schwarz, I made an effort to do what I could, and the book under the title "The Schrödinger Equation" appeared in 1983 in Russian. The translation into English took another 8 years, although work on it began immediately (the English version was published by Kluwer in 1991).

Another period in my life in which Berezin played a very important role came around 1974, when I decided to try to submit a Doctor of Science thesis. I had done some things about spectral theory of almost periodic differential operators and in particular found a way to apply von Neumann algebras there. Everyone around me said that I had gathered enough material for this thesis. All except Berezin. He explained a few things about it to me. He said that my results lack applications to mathematical physics and suggested a direction for getting these applications. He called my attention to solid state physics and physics of disordered systems and conjectured that the integrated density of states should coincide with the spectrum distribution function that I had defined with the help of von Neumann algebras. Eventually I proved this conjecture and it became one of the best parts of my thesis. Berezin also insisted that the main results should be published in the Russian Mathematical Surveys and reported at the Moscow Mathematical Society. I did all this, although at the moment I was reluctant about it because this caused a considerable delay (and I knew perfectly well that theses of much lower quality were defended without such additional precautions). But later on I had a chance to see how wise Berezin was in giving me his advice. Shortly after I submitted my thesis (already satisfying all Berezin's requirements), certain people started trouble, trying to topple me. For me it was a fight for survival and I do not know what could have happened if I had lost it. But I won in the end, because of support of such mathematicians as V. Arnold, L. Faddeev, A. Kolmogorov, V. Marchenko, V. Maslov, S. Novikov, S. Sobolev, support that they could hardly have provided if my thesis had been as I had planned it before talking with Berezin.

Now it is obvious that the name of Berezin will be never forgotten in mathematics and mathematical physics (if only because of the Berezinian or of the Berezin integral over anticommuting variables). But it is important to remember that he was also a man of the highest moral standards, and besides, was very kind and warmhearted. This helped people who learned from him, worked with him, and just surrounded him. Many Moscow students who were simply in his classes remember him as one of the teachers who regarded students as their equals (and to my understanding, such teachers were not a majority in Moscow then). I personally know that my meeting with him in 1967 changed my life and am infinitely grateful to Berezin for this.

MY RECOLLECTIONS ON BEREZIN *

A. M. VERSHIK
Mathematical Institute of the Russian Academy of Sciences
27 Fontanka, 191011 St. Petersburg, Russia
vershik@pdmi.ras.ru

Berezin's departure from this world was unexpected and mysterious. There is an element of mystery in any death, but Alik's death, far away from Moscow, in a geological party, where there were no friends, under circumstances that remain unclear, was a strange kind of disappearance. I remember the moment when I was told about it, and the overwhelming feeling of disbelief ...

Our last long conversation occurred half a year before that, at a summer workshop near Minsk. This conversation made a strong impression on me: I recognized Alik's acute pessimism. The year was 1979, we were deep in the "stagnation period" with its oppressive atmosphere, the absence of any hope for the liberalization of our society, vicious attacks against any form of dissent, active emigration, dismal days at the university. We strolled for a long time in a quiet forest. Alik touched upon the usual topics of our discussions-the situation at mekh-mat, on the hopelessness of trying to improve it, on the problem of mutual friends who were emigrating or weren't, on the impossibility of real contacts with Western mathematicians, and how this fact is exploited by some both here and abroad. But the main topic of our talk that evening was the Jewish problem, which we had rarely discussed before. I recall that it was this part of the conversation that struck me most of all; such gloomy forecasts

* Translated from Russian by A. Sossinsky, Independent University of Moscow, Bolshoy Vlasyevskiy Pereulok 11, 119002 Moscow, Russia; e-mail: lifrc@mccme.ru. Reprinted from *American Mathematical Society Translations*, Series 2, vol. 175, pp. 228-231, 1996, with the kind permission of the Publisher.

of events to come I had rarely heard from anyone before; Alik was saying that he was afraid of pogroms and open persecution, that communofascist ideas were in the air, and so on. To the workshop I had taken with me some samizdat and tamizdat materials and our journal Summa (a Leningrad samizdat publication, surveying a wide spectrum of political, social, and literary questions), where Jewish problem was also discussed; I had shown all this to Alik. Like some authors from the dissidence movement, I viewed the future dangers under a different angle, and tried to convince Alik that pogroms were hardly an imminent reality. Today, after all these years, one can say that Berezin's predictions were, several times, on the verge of coming true, his intuition did not entirely deceive him. However, later, coming back to that conversation, I always felt that Alik's apocalyptic vision, in some mystic sense, was not an accident.

I first me Berezin in the early sixties here in Leningrad; he was at the height of his popularity was often invited to our city. My first conversation with him disconcerted me somewhat: he said that my results on Gaussian dynamical systems had been known for a long time, but then agreed with my objections; I appreciated his ability to immediately get at the core of any subject and, of course, his wide erudition. Since then we met many times and talked about mathematics, about friends, and acquaintances. These encounters did not occur too often, maybe two or three times a year, and I mostly listened and asked questions, at least in the first years. I think that it was progressively that an inner contact developed between us, the inner contact commonly known as friendship. It was easy to understand why close interaction with him was not easy to achieve: Alik often spoke with an aplomb that allowed no questions, was sometimes superficial, but these were minor traits in his overall intellectual image.

We in Leningrad often invited him for talks at sessions of the Leningrad Mathematical Society, at the V.I. Smirnov–O. A. Ladyzhenskaya seminar, and elsewhere. It seems that he enjoyed coming to visit us and engaging in discussions with colleagues whose work was within his sphere of interest (L. D. Faddeev, O. A. Ladyzhenskaya, M. S. Birman, V. S. Buslaev, and others). I participated in many workshops together with him (Katsiveli 1966, Kazan 1971, Tashkent 1975, Minsk 1979, among others). Always his participation was active and significant. His desire and aptitude to adapt to new interesting developments in mathematics that were presented at conferences and workshops was remarkable: he always strived to find a place for them in his continuously reconstructed physico-mathematical universum. I remember how early he had pointed out the role of topology in mathematical physics. On the other hand, he did not hide his attitude to pseudoscientific or superficial publications or reports, especially in areas adjoining physics.

Berezin occupied a special position in the Moscow and All-Union mathematical scene. He began as one of the most successful and cherished pupils of I. M. Gelfand in representation theory. The huge number of facts, inventions, unexpected relationships obtained at the time (the end of the fifties) constituted the core of this new theory: the role of Gelfand himself in this process is difficult to overestimate. However, many of these achievements were then only sketched, others required corrections and additions. This is also true of Berezin's first papers, written together with or under the guidance of his teacher. The shortcomings of these papers, or perhaps other circumstances resulted, unfortunately in the end of the cooperation between teacher and pupil, which occurred rather quickly, and this cooperation was never resumed again. Working with Gelfand in the seventies, and being friends with Alik, I tried to convince both of them in the usefulness and importance of their reunification, but was not very successful.

In attempting to assess his role in our mathematics, I should begin by saying that, in my opinion, it is precisely Berezin who was at the origin of the essential turning point in the work of many mathematicians and their rapprochement with physics. He was the first in his generation who, following his understanding of science, decided to carry out the huge efforts necessary to go into theoretical physics as a physicist rather than only as a mathematician, and succeeded in doing that. One can discuss at length to what extent this can or should be done, while still remaining a mathematician; there are examples when mathematicians became physicists, but Alik found his own proportion and became an active mathematical physicist and a propagandist of physical problems. He is the one who introduced many mathematicians to this circle of ideas, and some of them became outstanding experts in the theory of mathematical models of contemporary physics. Several participate in the present collection. The attitude of the leading physicists in the thirties, forties, and fifties to mathematics, although some of them were able to use the mathematical techniques of the times, was to say the least quite reserved. The fact that this had changed radically in the seventies, eighties and nineties should to a great extent be credited to Berezin.

Berezin's favorite topics were quantization and Grassmannian analysis. In the creation of Grassmannian analysis and supermathematics, he played the leading role. Essentially the general program for the construction of this theory was sketched in his first book (incidentally, this book, first written in the form of an article, remained an unsolicited manuscript for several years with the editors of Uspekhi, and was finally published in the series Biblioteka Uspekhov). For many years, he presented these ideas with great enthusiasm and tired to convince many people to work on them, but the real explosion of interest came much later.

Unfortunately, Alik only lived to see the very beginning of the unquestioned acceptance of his ideas. Today supermathematics (the term is accidental and imperfect, but that is no fault of Berezin's) has become a kind of parallel mathematics: any result must have its "superanalog."

Berezin's contributions to the mathematical theory of quantization are so varied that it is hardly possible to sketch them briefly here. I will mention only a few fragments. The popular notion of a quantum group is the development of the idea of deformation of the universal enveloping algebra put forward by Berezin, although in a form somewhat different from the one considered in the eighties. The history of the Lie–Berezin–Kirillov–Kostant bracket, which also appeared in connection with quantization, is well known. My impression is that Berezin himself regarded his cycle of papers on quantization as his main favorite theme.

I recollect many conversations with him on various topics. Most of them, in one way or another, had to do with different outgrowth of the main theme and can serve as examples of the application of physical ideas to purely mathematical problems. One of them (approximation in dynamical systems) was widely developed, although Berezin's role remained hidden in the background. He was interested in path integrals, where he justly considered himself to be one of the initiators, in the theory of von Neumann factors, C^*-algebras, asymptotic problems in algebra, and other topics. All the variations of the spectral theory of operators, scattering theory, the theory of matrix spectra, all this was always in the center of his attention. I remember discussions about the calculus of variations, on nonholonomic mechanics, on the algebraic aspects of the theory of integrable systems.

The heritage of a mathematician is never limited to his published work and even to his manuscripts; a part of it, usually difficult to perceive for future generations, is transmitted through reports, talks and conversations, in the ideas talked about with colleagues, and finally by the influence on others. All these components were strongly represented in Berezin's scientific life: by his talent and enthusiasm, his work, seminars, numerous reports and contacts, he succeeded in getting mathematicians, both well-known and young, interested in new problems.

His life in research, rich in scientific events, came to an early end. The hardships of existence in the Soviet Union for a talented scientist, who was not very loyal to the establishment and, in addition, a Jew, left a deep trace in Alik. I would not like to list here all the injustices and blows that he suffered from those in power, from his illwishers. Courageous by nature, Berezin always found the strength to rise above everyday annoyances and work, work. Was his outstanding talent fully realized? Can such a question be answered? Be that as it may, I feel he succeeded in telling us a great deal.

1996, St. Petersburg

ABOUT ALIK BEREZIN AND SOME OF HIS WORKS. TEN YEARS LATER *

A.M. VERSHIK

St.Petersburg Department of the Steklov Mathematical Institute,
27 Fontanka, 191011 St. Petersburg, Russia
vershik@pdmi.ras.ru

This paper is about my very good friend Felix Berezin and his mathematical works. In particular, I discuss the connection of his less known paper on projective representations of the current groups with my paper with I.M. Gelfand and M.I. Graev on multiplicative integral of representations. Ten years ago I wrote my first "Berezin note." It will be reprinted in this Collection. However, when I looked at it now I found many things about Alik which I did not mention but which should have been mentioned. The reader will find some of them below, while some others are left for future publications. Recently a memorial volume dedicated to Berezin's 75-anniversary was released in St. Petersburg. It contains mathematical papers of his friends and followers and was published in 2006 in Russian as Zapiski Nauchnykh Seminarov POMI, v. 331: Theory of Representations, Dynamical Systems and Combinatorics, XIV, Ed. A. Vershik. The English version will become available as an issue of Journal of Mathematical Science, v. 141, February 2007 (Springer-Verlag, Berlin). The text below coincides in part with my preface in that Volume.

This year, the remarkable man and mathematician Felix ("Alik") Berezin (1931–1980) would be 75. His tragic and untimely death did not allow him to see the great influence of his works and ideas on the modern mathematics and mathematical physics. Neither had he occasion to take advantage of the present openness of boundaries and the possibility to communicate with the whole scientific world, the possibility which he awaited but in which almost did not believe, and

* Translated from Russian by Natalia Tsilevich, St.Petersburg Department of Steklov Institute of Mathematics, e-mail: natalia@pdmi.ras.ru.

the absence of which made him suffer very much. His scientific fate, as well as the fates of many scientists of the Soviet period, did not meet his talent and the results of his work. In addition to obvious and purely Soviet obstacles that hampered his way, there were also others. His independence has particularity made him "inconvenient" for any authorities and even for some colleagues. Although he had a great number of pupils, headed a popular seminar, and was quite widely known in the country, he was, by nature, rather a "lonely wolf" and a "maitre" than a group leader.

The beginning was very impressive and successful. He was the favorite pupil of I. M. Gelfand; together with R. A. Minlos they were the first of the young successors of Gelfand in representation theory whose development essentially started only in the early 50s. I remember how a little bit later, at a conference on functional analysis, I.M. praised Berezin's works and was proud of him. They made together a number of important works, but then something happened and the relations were radically spoiled. Most of their close friends believed that the rupture was caused by an error in their joint work. Indeed, one of their papers contained an error, or even several ones, but it was an instructive error, one of those errors that constitute an indispensable and even useful component of scientific progress. There are a lot of examples of this kind.

Much later, in the 70s, when I was friends with Alik and actively collaborated with I.M., I spoke openly to both, and neither of them believed that this error was the real cause of the rupture; quite possibly, it only provoked the outburst of Gelfand's displeasure and irritation; at any case, this was what they both said to me. I repeatedly tried to persuade each of them to resume the relationship, which would be to the general advantage. Unfortunately, both, in almost the same words, said of their reluctance to resume the relationship and even essential impossibility of any contacts.

For Berezin, this rupture meant a great deal: instead of receiving support in his projects, which was badly needed in his case, he was confronted by misunderstanding and scepticism. His paper on Grassmann analysis, written in the early 60s, in which he laid foundations of what is now called supermathematics, was submitted to

Uspekhi Matematicheskikh Nauk,[a] but had not been published for a long time, and finally, after many years delay, appeared as a separate volume (maybe, fortunately). As to his "Doctor of Science" degree, which should be defended about ten years earlier, or his position at the Department of Mechanics and Mathematics (he never become a professor), here the causes were general, a whole circle of scientists were victims of this regime. Now it is difficult to believe that a good few of Soviet mathematicians, now world-famous, for years could not submit and defend a dissertation, obtain a title, etc.

The role of F. A. Berezin in the mathematical community of those years is extremely important. In short, it was he who, having profoundly studied theoretical physics (quantum field theory of the time, statistical physics, etc.), *was the first of his generation of mathematicians to popularize it and systematically engage active mathematicians from Moscow, Leningrad, and other centers in working in this field.* Almost all my friends mathematicians close to mathematical physics, from my and adjacent generations, had been some time under his strong influence (L. D. Faddeev, R. A. Minlos, R. L. Dobrushin, Ya. G. Sinai, A. S. Schwarz, and many theoretical physicists). His seminar of 60s–70s attracted dozens of mathematicians and theoretical physicists from the whole country. This should be described in detail by those who directly experienced this influence at that time. I know only the memoirs by R. A. Minlos and A. S. Schwarz in which they write about this role of F. A. Berezin.

One should keep in mind that the relations between mathematicians and theoretical physicists in the pre-war years and the first post-war decades were totally different from the present ones. The turning point happened in the 60s, and Berezin's role here was very significant. It is well known that even earlier I. M. Gelfand was very interested in physics, wrote papers on physical subjects, and, as well as Alik, at one time attended Landau's seminar. But the consequences of this interest in his own school did not advance beyond quite general knowledge of the state of art. I.M. once reproduced to me a very apt expression due to A. B. Migdal, that he used to

[a] *Russian Mathematical Surveys.*

go to physicists "as Narodniks to muzhik" (in other words, from the one hand, out of respect and wishing to learn a little bit, but, on the other hand, wanting to teach them reading and writing). I do not know to what extent this *mot* is exact, but certainly Alik mainly learned and "became a trained physicist," like very few professional mathematicians. He used to say "When I think as a physicist...". He was very respectful towards physics, scoffed at mathematicians who "wanted to finish physics," and believed that this is hardly possible.

My first acquaintance with Alik happened in the early 60s. He often visited Leningrad and liked these visits. He was friends with O. A. Ladyzhenskaya, had close scientific contacts with M. S. Birman and L. D. Faddeev. Later I often invited him to give a talk at our Mathematical Society. During our first discussion, in 1963 or 1964, we spoke about the Fock space, which I had just discovered for myself. As a matter of fact, in my Ph.D. thesis I, on the initiative of V. A. Rokhlin, studied the structures of the L^2 space over an infinite-dimensional Gaussian measure; in particular, V.A. asked me to explicitly determine in this space the structure of a unitary ring, in order to study spectra and other invariants of Gaussian ("normal") dynamical systems, etc. At the time, the interest to normal dynamical systems and Gaussian measures – their geometry, spectra, and other properties – was universal. The ideas of A. N. Kolmogorov, the works by S. V. Fomin, I. V. Girsanov, and, especially, the famous works by K. Itô – that was all that I knew on this subject at the time. But neither any of us nor the authors of these works knew about the link with the Fock space. Almost at the same time, there appeared N. Wiener's lectures on the nonlinear theory of random processes, which contained a description of the space of nonlinear functionals with respect to the Gaussian (Wiener) measure (by the way, not without essential errors); but in this book there was no word on second quantization. I remember that A. N. Kolmogorov, at a conference in Vilnius, asked me what was new in Wiener's book. L. D. Faddeev was the first to explain to me and Ya. G. Sinai what are "Fock columns." I dashed to read V. A. Fock's papers of 1932 in which he formulated the ideas of second quantization, and since then I regard this construction as one of the most important achievements of mathematical physics of the 20th century. And then, after I had read Fock's papers, my first encounter with F. A. Berezin happened. My own modest contribution to this problem was a formula for the multiplication of multidimensional Hermite polynomials, which allowed one to determine the structure of a unitary ring, obtain expansions of functionals in terms of generalized Hermite polynomials, etc., which can in turn be used in the analysis of normal systems and the corresponding operators. Now this space, which bears many names (Fock–Wiener–Itô–Cook–...), is generally known; it is the subject of several books (e.g., that by B. Simon). F. A. Berezin also wrote much about the Fock space, but he preferred the analytical (Bargmann) model rather than the probabilistic one.

After that first discussion on the Fock space, we used to meet each time when I visited Moscow or he visited Leningrad and discuss various mathematical and nonmathematical subjects. In particular, we both were interested in von Neumann's constructions of factors, C^*- and W^*-algebras, representations of the commutation and anticommutation relations, approximations (now, few people remember that it was Berezin who initiated the approximation theory of dynamical systems, which was successfully developed in the 60s by his pupil A. Stepin and Ya. Sinai's pupil A. Katok), and especially the theory of measures and integration in functional spaces, which at a time was his favorite subject and which then interested me very much too. I remember our long conversations in Katsiveli (Crimea) at a summer school in 1966, in which he developed his understanding of approximations and the theory of factors; and especially those in Chimgan in 1975, in Zvenigorod, and in the environs of Minsk in 1979, when, during our, alas, last meeting, we spoke about representations of current groups (which will be discussed in detail below) and the situation in science and society, about which I wrote in [10]. He supported very actively mathematical youth and spent lot of time for pedagogical reasons. I remember his story about D.Kazdan and others whom he helped during their early stage in mathematics. Son of my cousin physicist Dima Gitman whom I recommended to Alik obtained a great support from him when Gitman had needed in it very much (see his recollection in this volume).

One of the mathematical subjects most important for Berezin was the *mathematical theory of quantization*; he worked in this field, studying it from different viewpoints, continuously, one might say, all his life. F. A. Berezin made an invaluable contribution to the mathematical theory of quantization. I hope that some more competent specialists will describe in detail Berezin's approach to quantization and its evolution. This circle of problems includes geometric quantization (quantum mechanics); the orbit method, to the development of which he was directly related; deformation quantization, represented by the famous series of papers in which he tried to involve into consideration differential equations and the theory of complex manifolds; second quantization and the theory of the Fock space and Bogolyubov transforms; and, finally, a purely algebraic approach to quantization (joint works with the Kiev mathematician G. I. Kats). And, of course, his favorite creation – Grassmann theory and supermathematics – is directly related to quantization.

Here I will dwell on only one topic that interested us in the 70s. Namely, I want to tell about a problem that we studied together with I. M. Gelfand and M. I. Graev, and on which Alik also worked, the problem concerning representations of current groups, or the continuous tensor product of representations. He had arrived at this problem also from quantization, and regarded it as being close to the series of papers mentioned above. He dealt with this problem in two papers: "Representations of the continuous direct product of universal coverings of the group of motions of the complex ball" [1] in *Proceedings of the Moscow Mathematical Society* and its English version [2] in *Reports on Mathematical Physics.*

I must begin with a short history of our joint studies with I. M. Gelfand and M. I. Graev devoted to so-called integrals of representations or current groups. In spring 1972, at the end of a rather long visit to Moscow, I met I.M. and told him about various problems I worked on. He seemed to be interested. I spoke about the asymptotic statistics of permutations, dynamical systems, and, perhaps, C^*-algebras; there was no mention of representation theory, though at the time I intensively studied it. Having taken leave of I.M., I was going to return to Leningrad, but suddenly, at the last moment, he found me, in a rather complicated way, at my Moscow friends' apartment and asked me to come. When I arrived (I.M. had also invited M. I. Graev), he told me about a problem that, according to his words, he had been suggesting to many people for a long time (however, in view of the above, he hardly had discussed it with Berezin). The problem was as follows: does there exist an irreducible representation of the group of functions on a manifold with values, say, in the group $\mathrm{SL}(2, R)$ with pointwise multiplication ("the group of currents with values in $\mathrm{SL}(2, R)$") that is invariant with respect to some group acting transitively on the manifold (so-called "multiplicative integral of representations"). For example, does there exist an irreducible representation of the group of currents on the circle invariant with respect to rotations? In six months, this question was solved positively, and in January 1973, I send to I. M. Gelfand a letter containing a crucial argument; this was the beginning of our collaboration with him and M. I. Graev, which resulted in a long series of publications devoted to this and close subjects.

In December 1972, Alik came to Leningrad; he was invited to give a talk at the Mathematical Society and the Ladyzhenskaya–Smirnov seminar. We had long discussions; in particular, I told him about this problem. Just then I was preparing a letter to I. M. Gelfand with the calculation of what was later called the canonical state. Berezin became very animated and said that he also had thought about this problem, but in a quite different way. After a time, he wrote two papers on this subject; they were published much later, in the late 70s. Afterwards we did not essentially discuss them. At the time, I did not try to understand these papers, because my interest to this topic had already gone. They contain a formal reference to our work, but without any comments. Unfortunately, here also he had some feeling of resentment, which I tried to remove, but, alas, his complicated relationship with I.M. again played its role. This is not the right place to discuss these matters, so I proceed to the mathematical aspect of the problem.

It happened that recently, after quite a long time, we returned to these problems together with M. I. Graev, having in mind a number of new ideas, and I decided that I must eventually understand

Berezin's contribution to this subject and its relation to our old works. With a significant delay, I managed to straighten this out.

For the semisimple Lie groups $\mathrm{SU}(n,1)$, $\mathrm{O}(n,1)$, $n = 1, 2, \ldots$, the only ones for which such cohomologies do exist and were calculated explicitly in [4], it became possible to carry out the construction of multiplicative integral. Berezin in [1, 2] considered part of these groups, the universal coverings of the group of motions of the complex ball or the complex Lobachevsky space. Now, having looked closely at Berezin's paper [1], which is rather difficult to read, I came to the conclusion that, though Berezin had not written the cocycle explicitly, he had done the same work at a concrete level. Namely, he had directly constructed the required realization of the Fock space as the exponential space with respect to the integral of special representations of the group $SU(n,1)$; and the cocycle in his construction appeared automatically, due to his method of introducing the norm. However, the proof that the representation is irreducible in his approach is more routine, while in our setting it follows from a general theorem on the canonical state. Moreover, it turned out that the notion of canonical state, introduced in [3], was essentially used by Berezin, but with another definition, and later it was the subject of many papers: e.g., by V. F. Molchanov (F. A. Berezin's pupil), G. van Dijk, and their successors.

Thus F. A. Berezin suggested his own independent construction of multiplicative integral of representations for the group of motions of the complex Lobachevsky space, which leads to one of the constructions of [3, 4], namely, the Fock construction.

Let me mention in passing that later, in the early 80s, there appeared quite new constructions, the so-called commutative models of multiplicative integral. First, in [8], this was done for $\mathrm{SL}(2, R)$ (see also [7, 12], where the commutative model for $\mathrm{SL}(2, R)$ is considered from the probabilistic point of view). And now, in our recent papers [5, 6, 9], we consider these questions anew and in detail. In particular, we study the commutative model for the general case $\mathrm{SO}(n,1)$, $\mathrm{SU}(n,1)$, and reduce the construction of the required representation to the construction of a representation of the maximal parabolic subgroup. For the time being, this is done for the group itself, but not for the current group. This trick significantly simplifies the construction and leads to quite new connections (random Lévy processes, infinite-dimensional Lebesgue measures, etc.).

Returning to Berezin's papers, I want to illustrate by this example the characteristic features of Alik's work, though I do not presume that my observations are of general nature. First of all, and this unites him with his adviser, he was concrete and not very fond of general constructions, though he knew that sometimes they might be very useful. Second, he had a remarkable intuition, which also unites him with I.M. In the example we discussed above, this intuition showed itself especially strong, because he had not a general picture of multiplicative integrals and did not know that the existence of continuous tensor products imposes very special conditions on the group and its representation. But if he did not know the corresponding theorem, and this was most likely the case, then he had to guess that rather heavy calculations would eventually lead to success. However, his skill in calculations and persistence were well known. But perhaps the most important feature of Berezin's work was his ability to see the perspective and persevere in carrying out a chosen program. His program of Grassmann analysis and supermathematics is one of the best manifestations of his insight.

References

1. F. A. Berezin, "Representations of the continuous direct product of universal coverings of the group of motions of the complex ball," *Trudy Mosk. Mat. Obsch.*, **36**, 275–293 (1978).
2. F. A. Berezin, "Representations of the infinite direct product of universal coverings of isometry groups of the complex ball," *Rep. Mathematical Phys.*, **9**, No. 1, 15–30 (1976).
3. A. M. Vershik, I. M. Gelfand, and M. I. Graev, "Representations of the group $SL(2, R)$, where R is a ring of functions," *Uspekhi Mat. Nauk*, **28**, No. 5, 83–128 (1973).
4. A. M. Vershik, I. M. Gelfand, and M. I. Graev, "Irreducible representations of the group G^X and cohomology," *Funct. Anal. Prilozh.*, **8**, No. 3, 67–68 (1974).
5. A. M. Vershik and M. I. Graev, "A commutative model of a representation of the group $O(n,1)^X$ and the generalized Lebesgue measure in the space of distributions," *Funct. Anal. Prilozh.*, **39**, No. 2, 1–12 (2005).
6. A. M. Vershik and M. I. Graev, "The structure of the complementary series and the special representations of the groups $O(n,1)$ and $U(n,1)$," *Uspekhi Mat. Nauk*, to appear.
7. A. M. Vershik, M. Yor, and N. V. Tsilevich, "On the Markov–Krein identity and quasiinvariance of the gamma process," *Zap. Nauchn. Semin. POMI*, **283**, 21–36 (2001).
8. I. M. Gelfand, M. I. Graev, and A. M. Vershik, "Models of representations of current groups," in: *Representations of Lie Groups and Lie Algebras*, A. A. Kirillov (ed.), Akadémiai Kiadó, Budapest (1985), pp. 121–179.
9. M. I. Graev and A. M. Vershik, "The basic representation of the current group $O(n,1)^X$ in the L^2 space over the generalized Lebesgue measure," *Indag. Math.*, **16**, No. 3/4 (2005).
10. V. P. Maslov, M. A. Shubin, A. M. Vershik, and N. D. Vvedenskaya, "Alik Berezin in the recollections of friends," *Amer. Math. Soc. Transl. Ser.* 2 , **175**, 225–236 (1996).
11. Y. Shalom, "Rigidity, unitary representations of semisimple groups, and fundamental groups of manifolds with rank one transformation group," *Ann. Math.*, **152**, 113–182 (2000).
12. N. Tsilevich, A. Vershik, and M. Yor, "An infinite-dimensional analogue of the Lebesgue measure and distinguished properties of the gamma process," *J. Funct. Anal.*, **185**, 274–296 (2001).

REMINISCENCES OF A CLOSE FRIEND *

NIKITA VVEDENSKAYA
Institute of Information Transmission Problems
Russian Academy of Science
19 Bolshoi Karetnyi
Moscow 101447, Russia
ndv@iitp.ru

These notes make no claim to completeness and no attempts at any generalizations. They are simply the reminiscences of a close friend.

I met Felix Aleksandrovich Berezin in 1948, when we were both admitted to the Department of Mechanics and Mathematics (mekhmat) of Moscow State University and were placed in the same study group. [a] We all called him Alik (this was the name used by his family and his close acquaintances), and I will use this name here. By the time we graduated, we had become close friends, but we got along quite well from the first year. I learned about his family and home later from his words.

Alik was brought up mainly by his mother's parents, because his mother herself, Esther Abramovna Rabinovich, was too preoccupied with her career (she was divorced from Alik's father). Alik recalled that in her youth she had envisioned the career of a concert pianist, but decided to become a doctor instead for ethical reasons. When

* Translated from Russian by A. Sossinsky, Independent University of Moscow, Bolshoy Vlasyevskiy Pereulok 11, 119002 Moscow, Russia; e-mail: lifrc@mccme.ru. Reprinted from *American Mathematical Society Translations*, Series 2, vol. 175, pp. 231-236, 1996, with the kind permission of the Publisher.

[a] Students in Soviet universities and present-day Russia normally have the same prescribed curriculum and are organized into "study groups" of 20–30 people. All the groups attend lecture courses together. They have separate problem solving sessions, however. Each group is often a closeknit organizational unit, not only in the administrative sense. –Translator's note

he obtained his passport, Alik chose his mother's "nationality," officially becoming Jewish, and this choice was to play a crucial role in his life.[b] His mother's family came from Moldavia. Her brother as a very young man had enlisted in Yakir's Bolshevik troops, becoming a soldier of the Red Army. Later he went on to be an important industrial administrator and, as such, perished during the Great Terror in the late 1930's. (Incidentally, Yakir's name was always pronounced with respect in the family.) Alik's mother was not arrested because she left Moscow at the time (a timely departure would sometimes save people from arrest, especially if they were not the "principal candidates for arrest").

I write about this because the dread of terror always hung as a dark shadow over the family.

In the beginning of 1953, during the peak of the campaign against "rootless cosmopolitans" and "The Case of the Doctors"[c] rumors began to circulate in Moscow about imminent arrests of Jews and their expulsion (the rumors were founded — Stalin did have such intentions). Alik and his mother, expecting arrest, prepared small suitcases with underwear and medical supplies (the experience of 1937!).

Unfortunately, this complex system of fears remained with Alik for the rest of his life. He believed that all of us were under constant surveillance and was inclined toward gloomy predictions. For example, before the Moscow Olympic Games (in 1980, the last year of his life) there was a rumor that during the games people would be allowed to come to Moscow only with special passes. Alik surmised

[b]In Russian the word "nationality" is used to mean ethnic origin. What is known as nationality in the West is called citizenship. Each Soviet citizen, upon receiving a passport at the age of sixteen, had to indicate his "nationality." If both parents had the same nationality his would be the one of the parents; if not, he could choose between the two nationalities of the parents. –Translator's note

[c]At the end of 1952, a number of medical doctors, the most accomplished medical specialists, were arrested on fabricated charges. All of them, with the exception of two, were Jews. They were accused of acting under the orders of an American Jewish "spy" organization, *Joint,* with intention to kill the leaders of the Communist Party and the Soviet Government under the pretense of healing them. It was only Stalin's death that saved them from capital punishment. Shortly after Stalin's death it was announced that all charges were fabricated. – Editor's note

that this restriction would remain for good.

We, people from Alik's circle of friends, often made fun of his fears, but he was never able to overcome his eerie visions and the anxiety acquired in childhood. (I am writing this in a completely different time; it is difficult, fortunately, for younger people to imagine the atmosphere of the Stalin and post-Stalin years. I hope that Alik's daughter Natasha will never experience such feelings.)

Having completed his high school education, Alik wanted to study at the Physics Department of Moscow University. He had graduated from school with a gold medal, which according to the existing rules gave him the right of having the entrance exams replaced by an interview. But in 1948 practically no ethnic Jews were accepted to the Physics Department; after an unsuccessful interview there, Alik succeeded in entering the Mechanics and Mathematics Department.

From our first year we became part of a circle of friends; among other members I should mention M. Agranovich, A. Yushkevich, S. Kamenomostskaya, O. Ziza, and V. Ryzhik. Usually we sat together at lectures. Later all of us would go ice-skating. Outwardly Alik was not particularly striking. He was always dressed neatly but modestly, not to say poorly (although, of course, most people dressed in that way then). Actually he had a snug jacket with warm lining. During the lectures (which were obligatory) we would sit near an open window. I would feel cold, and Alik would offer me his jacket. I spent most of the lectures in his jacket, while Alik remained in shirt-sleeves without any problems. I later had the opportunity of noting his immunity to cold during our long walks, which are described below. I recall how one very cold spring we were canoeing outside of Moscow and Alik's canoe companion paddled in such a way that Alik was wet from head to toe, but during the whole day Alik never said a word.

Before the beginning of studies in our first year, all freshmen students were sent to work on the construction of the new Moscow University building on Lenin Hills (we studied in the old building on Mokhovaya Street). Working at the construction site, students got to know each other. After that Alik was elected *komsorg* of our study group,[d] and this meant that he was popular among the students of his group; one should not forget that at the time practically all students were Komsomol members. Of course Alik never became a person making a political career, although he was a very noticeable figure in mekh-mat. Incidentally, at the end of our fifth year (our last year of studies), he edited a very nonstandard issue, with "purely imaginary" number i, of the wall newspaper of our graduating class.

Alik studied extremely well and was one of the best, if not the very best, student of our class. From the freshman year he participated in E. B. Dynkin's seminar, then started going to I. M. Gelfand's. Nevertheless, he was not even recommended for graduate studies (this was 1953, and he was Jewish), even though his "diploma thesis" was highly commended. The defense of the thesis took place at the Chair of Algebra, then headed by A. G. Kurosh.

After graduating from university, Alik began working as a school teacher: no other job was available for him at that time.

At the university he was on friendly terms with many people, but had few close friends. He himself regarded some two to three people as his close friends in his youth and four to five people in his mature years. In them he appreciated originality of mind, sense of humor, and a sense of purpose. Professionalism was also very important to him, but not understood in the narrow sense of professional success. Thus, in his friend Valery Nikolsky, a physicist, he mainly appreciated erudition, Valery's rebellious nature, and even his devil-may-care attitude and relish of boisterous festivities (Alik himself was not inclined to the latter). In himself, Alik mainly appreciated his professionalism as a mathematician.

Alik was an introverted person, and he was easily hurt. He suffered considerably from the cool reaction to his mathematical ideas and achievements. For this reason, I believe, he parted with several people for whom he felt a strong sympathy in his youth.

His faith in mathematics and in his studies was the main trait of his character. I recall that in some context (I suppose I was trying

[d]Komsomol leader of the group. *Komsomol* is the Russian language acronym for the Young Communist League. –Translator's note

to convince him to go somewhere to relax) he said sharply: "I have no private life other than mathematics" (this is when he had a wife and a daughter, and his mother was still living with him). Many times he refused to go on vacation, declined other entertainment, so as not to be distracted from his work. Here he was quite capable of not keeping a promise, or of letting down his friends with whom he had intended to spend some time.

In Moscow, while living in the city, Alik subjected everything to his working schedule and was very serious about keeping himself in good mathematical shape, which included taking care of his health and making sure he got enough sleep. (He even claimed that he became seriously ill as a result of one sleepless night.) One of his favorite sayings was "I love having a good time, especially getting some sleep." He was always very anxious about his health (which was excellent); worried a lot about it, and, the pessimist that he was, predicted his own early death (alas, this prediction came true).

Recreation and physical activities — he liked to move — were also very important for Alik. Recreation was ordinarily "to take a walk." In the evening after work he would stroll in the streets around the house; on weekends he would leave the city. These weekly hikes played an important role in his life. They were in fact the source of our friendship.

Alik knew the outskirts of Moscow quite well, and visited them often. On weekends he would hike by himself or more often with a few friends. Beginning from our last years at mekh-mat (1952–53), I started participating in these walks regularly, on foot in summer, on cross country skis in the winter. At different times various people from our graduation class took part in the hikes: we often hiked with A. Yushkevich and V. Nikolsky, less often with D. Kazhdan, A. Schwartz, E. Fradkin. The women were usually S. Kamenomostskaya, I. Karpova (Alik's first wife), and R. Kallosh. We would take a commuter train, and walk for 20–30 kilometers, sometimes more. There were some records being set, like hiking from Serpukhov to Kashira in one day, a distance of 60 kilometers. We would walk particularly without stopping, eat sandwiches, stopping only to swim if there was water on our way (Alik was fond of swimming and diving).

Occasionally we would make campfires. In the beginning we usually hiked around the Savyelovskaya railroad line (or walked from it to the Yaroslavlskaya or Leningradskaya line), because Alik lived near the Savyelovkii train station. The fact that I lived at the opposite end of the city was never taken into consideration, especially since the Northern outskirts of Moscow are reputed to be the most beautiful. When Alik moved closer to the university, we started hiking mostly near the Kievskaya rail line. We walked through forests and fields, often made new tracks in virgin snow in the winter. During the spring thaw we usually walked along the *betonkis*[e] encircling Moscow, for at that time there was practically no traffic on them. During his youth, in the spring and fall Alik would wear cheap army-type boots, which he considered very comfortable (for me these boots were always associated with soldiers on the march). While on skis, he always wore a flannel sports outfit of the type that was worn by everyone in the early 1950s. Only towards the end of his life did he acquire a "modern" skiing suit and a lovely ski cap. On the other hand, his skis were always of high quality. During our trips, Alik was always the one to choose the itinerary. When the two of us went for walks, somehow there was no incentive to argue, but when there were many of us, arguments did arise. But usually Alik would listen to no one, and walked where he saw fit, and the others obediently followed suit. He exhibited the same stubbornness in longer trips as well.

Camping trips that lasted several days, summer or winter excursions into the wild, constituted a very special part of our existence. It is impossible not to mention them while describing the lives of our friends and our own. We loved to travel; we enjoyed remaining alone with nature, and at that time travel within the country was relatively cheap. Usually in the summer we would not go to vacation resorts, nor to somebody's *dacha.*[f] We would go camping, to the mountains and the taiga by foot, to waterways by canoe. (Actually in the summer Alik would usually rent a dacha in order to be able to

[e]Concrete roads made for the military. –Translator's note
[f]Country cottage or cabin. –Editor's note

live and work there and have the opportunity to take little walks and go swimming). Several times Alik went on winter skiing trips and in the summer, on canoeing trips on the northern and Siberian rivers and lakes. He went to very few mountain trips, for I think he did not like or understand mountains. In the mountains also he preferred to "go his own way": avoiding paths progressively curling uphill, he would go straight up the steepest incline, and move downhill in whatever direction he saw fit. In a similar way, on a lake he might choose to go "straight out to the open sea" even in stormy weather. In other words, he was not an easy companion on such trips.

In the spring, during freshet, we often went canoeing for several days, usually during the May Day holidays. Alik was a strong paddler. He was never afraid of the cold (as I mentioned previously), was always an expert campfire maker at our stops, and could blow up a soccer ball with his mouth. In general he was quite strong, especially in his hands and fingers.

He was also fond of working with his hands, say to repair a damaged canoe or even to fix broken skis. Once, using part of a broken kayak and ski poles, he contrived a kitchen table of which he was very proud. In his apartment and in mine he drove nails into concrete walls and was indeed a jack-of-all-trades.

As I have already mentioned, Alik's general appearance was not especially striking. He dressed very modestly: he preferred informal checkered shirts and sweaters to a suit and a tie. At social gatherings he tended to remain quiet, although he was an excellent conversationalist. Many thought he was conceited, because he often failed to say hello to persons he met at the department, but this was due to his poor eyesight. He simply did not recognize people (he only wore glasses at lectures and during classes). He was unpretentious in his everyday needs, in particular concerning food. I think this was due to family traditions (in his youth his family lived rather poorly). When he began working at the university, Alik always had leftover money from his salary, which he readily loaned to people, not putting up a fuss if the money was not returned ("I understand that he has children and no money," he once said about one of his debtors).

On our walks together, or telephoning each other several times a week, we spoke of many things: about life and people around us, about politics, about books we had read. Conversing, we enjoyed evoking events from our student life or camping trips. Together with A. Yushkevich, the three of us sometimes decided in advance who would be the storyteller of one of our favorite episodes. In a more serious vein, I should mention that Alik's forecasts for the future of our country were not very cheerful (but who could predict the approaching disintegration of the Soviet system!). Nevertheless, he loved Moscow, the university, and his way of life. At different times he had different thoughts about emigration (they depended on his mood and the level of unpleasantness at the university). But once, during the peak of emigration, to my direct question about leaving the country he answered: "No. I love the countryside near Moscow. I love the Russian language. I have several friends here, you among them."

Like all of us, Alik actively desired to see the world, to visit various countries. Even more ardently, he wanted to travel for contacts with mathematicians and physicists. But he was deemed "unsuited for travel" by the administration. He was however, allowed to go to Poland and Hungary (inside-the-iron-curtain countries). In Hungary he first lived at Karoly Majus' place and later simply in an office of the Mathematics Institute (an example of his unpretentiousness). He also visited Mongolia. He gave interesting descriptions of all these trips. He also told me about his first trip to Kolyma (I vividly recall the picture he had drawn of the ecological disaster due to the effect of gold prospecting dredges, which had literally turned a river bed inside out.) During one of his vacations he got himself hired as a manual worker in a geological expedition. We sometimes took such trips instead of our traditional camping excursions; one of Alik's school friends, a geologist, worked in Magadan. The second trip to Kolyma was to cost Alik his life.

As I mentioned before, Alik wanted to see the world, especially so because in the 1970s he had numerous invitations from abroad. But trips abroad were then impossible. Alik wrote a letter to the Rector of Moscow University, R. V. Khokhlov, describing the appalling status of all the talented mathematicians and their students at mekh-mat. After Khokhlov's accident and death, this letter came into the possession of the department's party and the administrative leaders. They did all they could to "punish" Alik and not to allow him to travel anywhere. Even after the new rector, Logunov, had authorized his trip to Poland, the department let the affair drag on for over a year, shuffling the official papers back and forth between bureaucrats' offices. Alik died before the final permission was given. I would like to stress that despite his prevailing interest in mathematics, Alik was not indifferent to questions of social justice. The letter to Khokhlov is an example.

Alik had a very poor command of foreign languages. In the last years of his life, when physicists desiring to talk to him would come to Moscow, he asked me several times to participate in long strolls with them (during such strolls mathematics and physics were thoroughly discusses) in order to help translate to and from English.

But Russian Alik spoke beautifully, in a rich, sometimes somewhat bookish language. He read a great deal and had an excellent literary library (besides the mathematical one). He was a widely educated person, was fond of classical literature, history, and read with pleasure such things as Chinese chronicles. His favorite books were[g] *The Adventures of Brave Soldier Švejk*, *Penguin Island*, and the poems of Heinrich Heine. Alik also loved music. He did not go to concerts often, but his mother always played piano at home.

In the summer of 1980 Alik went on his second trip to Kolyma, again to work in a geological party. According to the account of the other workers in his party, they were going downstream in an inflatable boat on the day Alik was to leave camp and return to Moscow. These boats are always difficult to maneuver. They got caught in a strong current heading for a heap of logs. Such accumulations of logs in Siberian rivers can be extremely dangerous: the stream plunges under the logs that stick out like the needles of a huge hedgehog. Everyone jumped out of the boat, pulled it out (thus the water was

[g] *The Fateful Adventures of The Good Soldier Švejk* by Jaroslav Hašek, a Czech humorist and satirist (1883–1923) and *Penguin Island* by Anatole France (1844–1924), one of the major figures of French literature in the late 19th and early 20th centuries. Heinrich Heine (1797–1856) is one of the most significant German poets. –Editor's note

not deep), and when they looked around, Alik was not there. No one saw him getting carried under the logs. He drowned. His wife, Elena Grigorievna Karpel and V. P. Palomodov flew to Kolyma and brought back his remains to Moscow. Alik was survived by his four year old daughter.

As any actively working mathematician, Alik wanted his work to be recognized by his peers. Recognition came to him at the very end of his life, and to a greater extent only after his death.

I have not written about Alik as a mathematician since we did not intersect professionally. But we saw each other very often: he sometimes came to my parties; he would usually come to my birthday celebrations and give me roses. At my housewarming party, he presented me with a hammer with an inscription engraved on it. Very rarely, only a few times in all these years, did he organize parties at his home, inviting his students and friends. When I broke my leg, he visited me at the hospital. He was my close friend, a person I could always rely on.

Moscow, 1996

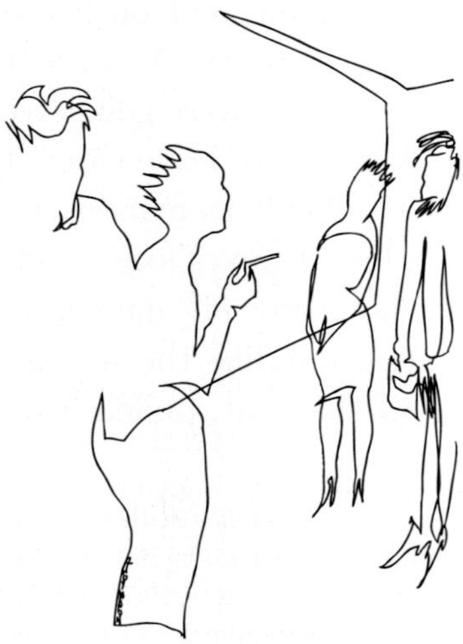

MY ENCOUNTERS WITH FELIX ALEKSANDROVICH BEREZIN *

SNAPSHOTS OF OUR LIFE IN THE 1960s, '70s AND BEYOND

DMITRI GITMAN

Instituto de Física
Universidade de São Paulo
Caixa Postal 66318-CEP, 05315-970 São Paulo, S.P.
Brazil
gitman@dfn.if.usp.br

Acquaintance *in Absentia*

The name Berezin first caught my eye on the cover of the book *The Method of Second Quantization*, that came on sale in 1965 at our Bookstore of Scientific and Technical Literature. I used to visit it almost every day in my fourth year as a student of theoretical physics at the Physics Department of Tomsk State University (which is in Siberia — for those who have never lived in Russia). By that time I had been deeply involved in quantum mechanics and quantum field theory, and the words "creation and annihilation operators" were music to my ear. I thought I was an expert in that area, after I had studied this operator technique as expounded by various authors, in particular, in Bogolyubov's textbook *The Method of Second Quantization*, written in Ukrainian and being then in circulation, and in Schweber's book *Introduction to Quantum Field Theory*, which I then considered a sort of Bible. However, right after a first superficial acquaintance with Berezin's book I realized that I had been actually

* Translated from Russian by Pavel Moshin, Iinstituto de Física, Universidade de São Paulo, Brazil; e-mail: moshin@dfn.if.usp.br.

acquainted only with the arithmetic of the subject, while I had yet to learn its algebra and analysis. Even though it was a few years later that I actually found it necessary to read and literally work my way through this remarkable book, Berezin the author, as any author of books on theoretical physics at that time, immediately came on my list of "celestials" (it was only much later that I finally realized that truly inspirational books written by experts were oh-so-rare. I always keep them all, and this book of Felix Aleksandrovich is the first one on my list).

It so happened that in the solution of most of the scientific problems I had to deal with since my student days, I always used, in some way or another, the results, ideas, and methods presented in the books of Felix Aleksandrovich Berezin. I was also lucky enough to meet him personally and even to discuss with him some problems of both scientific and non-scientific character. It must be said that my reminiscences of Berezin are automatically mixed with my memories of certain problems related to his ideas and methods, and in some mysterious and irresistible way bring along certain recollections, snapshots of our life in those days. Let the reader forgive me if I should go a bit far on those.

How we Wound up in Siberia

I am going to make a digression of personal nature, since I believe that the reader will find it easier to understand my feelings and perspectives at that time and, perhaps, at all the times to follow. I was born in 1944, in Tashkent, where my mother had arrived after her evacuation from Kiev. Her father (my grandfather), professor of philosophy, director of the National Ukrainian Library, was arrested in 1936 and sentenced to 10 years of imprisonment, without the right of correspondence,[a] to be followed by an exile to Kolyma (he was actually executed in 1937 in Moscow, as I was informed in Maga-

[a] A special clause in the cases of political victims of Stalin's purges. In fact, the sentence "10 years of labor camps without the right of correspondence" implied immediate execution of the convicted. –Editor's note

dan[b] in 1973 by a certain high-rank *kagebeshnik*,[c] an acquaintance of my acquaintances, who consoled me, saying, "your grandfather was really lucky that he never reached this place"). Her mother (my grandmother) was then in exile in the city of Ufa, as the wife of the "enemy of the people."

My father, having spent three years at the front line[d] and being heavily wounded, was demobilized and returned to Tashkent (the first day he had been discharged from hospital there, still bandaged all over, his wounds yet unhealed, he was walking through the city and some hoodlum attacked him, shouting, "while we fight the enemy, you Jews cosily wait through the war in hinterland").

When the war was over, our family moved to Leningrad, and the four of us (along with my grandmother, who had returned from exile) lodged in a room of a communal apartment; my grandmother, having a uniquely possessive character even for a Yiddishe mama, was a great complication in the life of my parents, according to my father's recollections. My father had lost his job at that time, and my mother worked as a lab assistant at a medical institute. As a result, our family broke apart, and my mother, a very beautiful woman, accepted the marriage proposal of a well-known professor of therapy (who had been Head Therapist of the Leningrad Front during the

[b]The Kolyma Region was one of the major locations for *gulag* camps during Stalin's rule. Magadan is the largest city of the Kolyma Region. *Gulag* is the Russian acronym for "The Chief Directorate of Corrective Labor Camps and Colonies" of the NKVD. It was the branch of the State Security that operated the penal system of forced labor camps and associated detention and transit camps and prisons. The *Gulag* system has become primarily known as a place for political prisoners and as a mechanism for repressing political opposition to the Soviet state. Though it imprisoned millions, the name became familiar in the West only with the publication of Aleksander Solzhenitsyn's *The Gulag Archipelago*, which likened the scattered camps to a chain of islands. –Translator's note

[c]KGB was the name of the secret police at that time. People often referred to the KGB agents as *kagebeshniks*. In Russian this word has negative, scornful connotations. The Soviet secret police changed acronyms like a chameleon. It started out as the Cheka, and then became the GPU, the OGPU, the NKVD, the NKGB, the MGB, and finally the KGB. It was quite common to refer to the secret police as "the Cheka" long after Cheka *per se* was no more. The secret agents were also called "the Chekists." Editor's note

[d]The author means WWII, which in the Soviet Union started with the invasion of Germans on June 22, 1941. –Editor's note

entire war, including the time of the blockade).

However, a short time later the professor was discharged from Leningrad Medical Institute after the beginning of the *Case of the Doctors*." [e] Some wise people advised him to leave Leningrad to avoid imminent arrest. So, we went to Siberia, to the fine city of Tomsk, where the then Rector of the Medical Institute had enough courage to employ my stepfather. Thus it came to pass that at the age of 8 I began my Siberian period of life that went on for 31 years. It was there that I finished school and, in 1966, graduated from the Physics Department of Tomsk State University. Despite the fact that I had been one of the best students (excuse me for the obvious lack of modesty, but what happened did happen), *magna cum laude*, I was not allowed to enter the graduate course at the University (oh, that sacred issue, "entry number five." [f] I didn't even complain, everybody knew what's what).

Nevertheless, as I can see it now, the system was rather rational; it did not throw capable people to the wind, and it had a way to use them reasonably well, though within some limitations. I was admitted as a graduate student to Tomsk Institute of Radio Electronics, under the supervision of Edward Abramovich Arinshtein, the then Head of the Physics Department, who had not been admitted to the University either. It was there that I started my PhD thesis on Quantum Statistics, whose basic tools were the technique of creation and annihilation operators and the method of a generating functional. Thus it naturally came to pass that I had to return to Berezin's book on second quantization. In 1969 I defended my PhD

[e] At the end of 1952 a number of medical doctors, the most accomplished medical specialists, were arrested on fabricated charges. All of them, with the exception of two, were Jews. They were accused of acting under the orders of an American Jewish "spy" organization, *Joint*, with intention to kill the leaders of the Communist Party and the Soviet Government under the pretense of healing them. It was only Stalin's death that saved them from capital punishment. Shortly after Stalin's death it was announced that all charges were fabricated. – Editor's note

[f] Soviet "passport" (which was in fact an internal ID card) carried extensive information such as home address, marital status, etc. Ethnicity — which in the Soviet parlance was referred to as "nationality" — was entry number 5. It was mandatory to disclose these data in all official documents. Entry number 5 became a euphemism for "Jewish." – Editor's note

thesis and in 1975 became Head of the Department of Mathematical Analysis at Tomsk Pedagogical Institute. There I was allowed (the system had decided that this was advantageous to the city) to create a scientific group (and even to accept two lads of "non-aboriginal" ethnicity, Joseph Buchbinder and Senya Shvartsman), which, in a certain way, exists up today. In 1985 I managed to get from Tomsk to Moscow, and since 1992 I live and work in São Paulo.

I Meet Berezin for the First Time

At that time we worked at the problem of the quantum creation of particles in strong external fields. That was the cause for my serious study of Berezin's book. It turns out that the key point in this kind of problems is the question of linear transformations of one set of creation and annihilation operators into another one. In Berezin's book such transformations were called canonical ones, and an exhaustive analysis of this and related questions was presented. Nevertheless, the term that has come into common usage is "Bogolyubov's transformations."

In fact, Bogolyubov did solve a number of problems by means of linear transformations for creation and annihilation operators; however, it was Felix Aleksandrovich who elevated this question from the level of "arithmetic" to a highly mathematical level, in particular, pointing out the importance of distinguishing between different types of transformations, "proper" and "improper." It was from Berezin's book that I learned that not just any transformation preserving canonical transformations was okay. It is only proper canonical transformations that retain the vacuum vector in the initial Hilbert space. Applying the criterion of "properness" established by Berezin to the transformations in my problem, I instantly found it to be equivalent to the clear physical condition of finiteness for the density of particles created from vacuum. I was overjoyed. When I told Felix Aleksandrovich about this fact later, I felt that he was pleased as well.

One of the problems I had been trying to solve at that time led me to a problem with a certain quadratic Hamiltonian. Following my experience and numerous examples from books and articles, I attempted to diagonalize this Hamiltonian, so as to bring it, in each mode, to the so-called canonical form $a^\dagger a$ which would solve the problem immediately. I was going to achieve this with the help of the already mentioned canonical transformations for creation and annihilation operators. I easily succeeded in my task for most of the values of a certain parameter that entered the Hamiltonian. Nevertheless, there remained some values of that parameter which, in certain modes, did not allow me to do so. Furthermore, it was clear that the spectrum of those modes was continuous, which implied that they could not be brought to the canonical form. I had not expected that.

Even in the book of Berezin this case had not been described explicitly, not to mention the standard textbooks on quantum mechanics. My discussions with the small circle of theoreticians available in Tomsk at that time failed to clarify the problem, and then I decided to go to Moscow to consult Berezin personally. I made a phone call to Anatoly Vershik, my first cousin once removed, a well-known mathematician, who lived in Leningrad and was on friendly terms with Berezin, asking him to arrange a consultation for me. When he did that, I arrived in Moscow and came to the Mekhmat of MGU,[g] where I was to meet with Felix Aleksandrovich.

Our meeting took place in one of the Mekhmat's corridors. As I learned afterwards, Berezin was not on a professor's salary; instead, he held a temporary position of a researcher and probably did not have a decent place of his own at the department's office.

At first glance I realized that he looked exactly as he should look like: a high forehead, a most clever and pleasant face of a very delicate and intellectual person. He was immersed in himself all the time, but I could see that he was making efforts not to create the impression that he did not find our conversation interesting, though why was he supposed to find it interesting to talk with an unknown young man from the provinces?

In an effort not to take too much of his time, I quickly explained

[g]Department of Mechanics and Mathematics of the Moscow State University. –Translator's note

the essence of my problem. He reacted immediately, without a trace of emotions, saying that in the general case a one-mode canonical transformation brings a quadratic form of creation and annihilation operators to one of the two canonical forms: one is $a^\dagger a$, described in every book, and the other is $a^\dagger a^\dagger + a\,a$. In addition, there exists no canonical transformation that brings one of these forms to the other.

I was dumbfounded, as I instantly realized that this fact solved my problem immediately. But how come it was not written anywhere (until now this simple yet nontrivial fact is almost unknown; I recently tested this on a person of a very high level)? At that moment, Felix Aleksandrovich, who until then had looked somewhat slow and distracted, suddenly turned around, livened up, and called a person passing by along the corridor,

"NN (I forgot the name), do you remember which are the representations of the group (I forgot which one)?"

It was obvious that after he had listened to the answer, one part of his mind was still occupied with our conversation, out of politeness, while the other part was already thinking over the response of that person. After that we said good-by, and I went away, at a slow pace, tormented by conflicting emotions and heading "home" (in those days, when I visited Moscow I stayed with Aunt Sonya, my mother's cousin who had a room in a communal apartment at Chistyie Prudy; her entire life she had worked as a cashier at a grocery store and barely made ends meet: a counter-example for anti-Semites, who believe that a poor Jew has never been born alive). I felt sad.

On the one hand, I realized that my campaign had been successful and that I had completely solved my problem. On the other hand, I had the same bitter thought, once again and not for the last time, that I was unlucky to live in the provinces and that I did not have the advantage of constant interaction with those who could really teach me a lot of things. My God, did I envy those chaps who lived in Moscow, visited scientific seminars, and could learn from really outstanding scientists, who were then numerous in the city. The problem was that in those days we did not have a free choice of the place of residence (the place of *propiska*[h]); it was only in 1985, when I was already a Doctor of Science,[i] that I managed, with the greatest difficulty (in fact, by cheating the system), to move to Moscow. *Perestroika*, that crushed the system in 1990, brought me to Brazil, once again making me a provincial, and once again I go to Moscow to rub shoulders with good people, among whom, unfortunately, Felix Aleksandrovich is no more.

A Seminar at the Institute of Physical Problems

In 1974 I became a member of the so-called Group of Efim Samoilovich Fradkin, who worked at the I.E. Tamm Theory Department of FIAN (the Institute of Physics of the Academy of Sciences of the USSR), and who invited me to work with him after I had made a report at a FIAN seminar. It should be noted that the first to invite me to FIAN seminars was Tolya Shabad, a man of nontrivial thinking. Now that I had started to frequent Moscow and become familiar with many theoretical physicists, I no longer was that provincial whom I had felt myself in the previous episode. Sometimes I would come across Felix Aleksandrovich at conferences and Moscow seminars. One of these episodes has deeply impressed itself in my memory. It was one of those seminars that took place at the Institute of Physical Problems, and was in fact a joint seminar of the Institute's Theory Department and of the L.D. Landau Institute of Theoretical Physics. The atmosphere of such seminars was quite peculiar (I believe it is familiar to many people, so my account here would not be the first one if I decided to go into the details). Let it be mentioned, however, for the sake of a better understanding of the episode, that the front rows of seats were always taken by Landau's direct disciples (and his former collaborators),

[h]A kind of a residence permit which, in effect, eliminated the freedom of movement inside the country. It was impossible for a non-Muscovite to settle in Moscow since the residence permit was never granted. –Editor's note

[i]The academic hierarchy in Russia follows the German rather than the Anglo-American pattern. An equivalent of PhD in the US is the so called *candidate* degree. The highest academic degree, doctoral (DSc), is analogous to the German *Habilitation*. – Editor's note.

then came the disciples' disciples and collaborators, and then, finally, the rest of the audience. In addition, the "big fish" who did not belong to the school of Landau (but who had been his friends) also seated themselves in the front rows. One could see that the distribution of the audience in the direction starting from the blackboard was very important there, in comparison, for instance, with FIAN seminars. That said, the atmosphere in the front rows was most unconstrained; all manner of topics, often non-scientific, would be discussed out loud, which continued during the entire seminar in case the speaker was not one of the direct disciples, or one of the privately recognized young stars. Besides, a speaker who did not belong to the latter categories would be continually interrupted, almost immediately after he began his report, by shrill shouts of the kind, "this is well-known since long ago", or "this is not the way to do it", or, most frequently, "all this is stuff and nonsense", with the shouter jumping out to the blackboard, pushing the speaker off, and making some extra improvisations. To me, the actual scientific hierarchy of all present was no longer a secret (doubtless, among them there were a lot of remarkable scientists), and as I tried hard to listen to the speaker I idly noticed the frequent discrepancy between the territorial location of a scientist and his actual scientific weight. All of a sudden, as I was taking a sympathetic view of the back rows of seats, I spotted Felix Aleksandrovich, who had settled down there with a modest look. My sensation of a dissonance was so immense that I decided to come closer and make sure it was actually him, Berezin. It was him, all right, and we exchanged some impressions; when the seminar was over, he left unnoticed and nobody called after him. I realized that most of those present did not know him personally and those who might have been familiar with him did not regard him as a scientist of the first magnitude which I believed he was, and, as we see now, he was in reality. This difference in the estimation of Berezin was symptomatic. Landau's school had its own perspective on how one should deal with physics, whereas, shall we say, the school of FIAN had a different perspective... . I believe that they discovered Berezin only after supersymmetry and superstrings had come into fashion, or perhaps after Sasha Polyakov had published his articles that contained Berezin's integral over anticommuting variables.

The Hassle with my Thesis Defense; Berezin is my Opponent

I am now about to relate some meetings with Felix Aleksandrovich and certain episodes that took place at the late 1970s. Once again, I have to make a digression to tell the reader about our life at that time, which may help understand some nuances of the following episodes that involve Berezin.

As I have said, at Tomsk Pedagogical Institute I was allowed to create a small group, based on people selected only on the basis of their scientific merits, without paying attention to somebody being "entry-five-invalid." This news reached FIAN, and so one day I was approached by Boris Voronov, a research associate of the Theory Department, and by Igor Tyutin, who was not an associate but who was a participant of the seminars and an insider of the Department. They told me the following story. It turned out that Tyutin, who did not have a position at FIAN, was working at a certain semi-classified technical institute, whose director, a well-known academician, had a hobby of theoretical physics. Being omnipotent, since directors of such institutions at that time had an absolute power and nearly unlimited opportunities, he had equipped that institute with a theory department (despite it did not exactly fit in the structure of the institute) that had been involved with purely theoretical physics, including quantum field theory and gravity. It was the head of that department (let me call him "K") who had employed Igor. It must be said that all who had ever met Tyutin, in some way or another, knew him to be a man with brains whose equal did not exist at that day and age, and the fact that he did not have a position at FIAN could only be explained by non-scientific reasons. Everybody put up with that fact, which I cannot fathom even now.

Tyutin himself was not greatly concerned by that state of things (or perhaps he made the best to conceal it). No matter what, by the time we had that conversation Tyutin had really got himself in trouble. The point was that he had already finished his DSc thesis, and still he had not been able to get "K" to provide him with some documents (I think it was a letter of reference from the *partbyuro*[j]) that were required for his defense.[k] It was clear that Tyutin needed to change his place of employment to get an opportunity to defend his DSc thesis. He had not been admitted to FIAN for certain reasons, some being objective (being Jewish on his mother's side) and some subjective (what if everybody could see who was the smartest one?)

So I was asked the following question: is there any chance Tyutin can get a job at Tomsk Pedagogical Institute, in such a way that he should not lose his *propiska* in Moscow, where he lives with his wife and son (once Moscow *propiska* was lost it could not be renewed)?

I got excited by that idea as it was clear that Tyutin's presence in our group would enable a fast transition from provincial physics to a totally different level of research. As I returned to Tomsk, I arranged a meeting with Limonov, the rector of my institute, and laid out the whole idea, telling him what a remarkable scientist Tyutin is and how we would all increase our research level while working by his side. More than that, there was a ready-made DSc thesis, so the institute would get a young doctor of science, which is of no small importance for the prestige of the institute.

Before I proceed with my story, I must tell the reader what kind of person Limonov was. He had come to the position of Rector from the *obkom*[l] where he had been responsible for the Science Department of Tomsk Region. If there were some "people normal in a certain sense" (those who did dirty tricks only out of necessity but not for the fun of it) who managed to survive in the system, Limonov was one of those. It was he who had employed me, and it was his wont to take liberties in the personnel policy. I think he did all those things simply because he wanted to show what kind of improvements he succeeded to make at a godforsaken pedagogical institute, planning to use that accomplishment to climb higher up the ladder of hierarchy. He was a polite and sensible person. I could see that what I had told him caught his attention and that he was making some mental

[j]The local committee of the Communist Party. –Translator's note

[k]That discord with "K" is quite another story, which seems ridiculous now but was not exactly ridiculous at that time, and I will not discuss it here.

[l]The Regional Committee of the Communist Party. –Translator's note

calculation. After a silence he asked me:

"How come such a brilliant scientist can't find a job in Moscow? Maybe he's a drunkard or a philanderer? Tell me straight. Being a philanderer is even a virtue in my opinion."

I could see that he was looking for an easy and rational explanation of the case at hand (the purely Russian name Tyutin could not possibly lead him onto another explanation) hoping that it is not the worst case scenario, namely, that Tyutin was a dissident, in which case there was nothing he could do about it.

And then, acting purely on intuition, I momentarily found a convincing explanation; I said:

"Victor Ivanovich, unfortunately Tyutin is not a philanderer; in fact he's an exemplary family man, it's just that his mother is Jewish; just take a look at his documents."

I could hardly express how happy and glad that Russian man was by the fact that someone's mother had turned out to be Jewish. Everything became clear; the danger was acceptable; the problem was solvable, and so he instantly said:

"I'm in."

That was how Tyutin got a job at Tomsk Pedagogical Institute and how we started working together on quantization, one of Berezin's favorite topics.

A short time later Tyutin defended his DSc thesis, and then I also finished my DSc thesis, whose defense was faced with some problems of its own. After all manner of discussions with Fradkin, it was decided to submit my thesis to the Scientific Council of Novosibirsk Institute of Nuclear Physics, where there were many experts on the problem of quantum electrodynamics with external fields. Vladimir Nikolayevich Baier, the group leader in the Theory Department of the Institute, gave his consent to be my opponent.[m]

[m]Thesis defense in the Soviet Union was a lengthy and complicated procedure. Dissertations ready for defense were sent to the so-called *official opponents* – experts from other institutions – who were supposed to study them carefully and then present their critical evaluations on the day of defense. Thorough debate was supposed to follow, and only after that the Scientific Council could approve (or decline) the dissertation under consideration. In the case of approval, dissertations were then sent to Moscow to the

Academician Budker, the then Director of the Institute, accepted the thesis for defense, but before he had time to register his decision officially he suddenly passed away. The new governing body of the Institute and those who had supported me there suddenly became over-cautious. They asked me to wait a little, as they had been expecting the move from Novosibirsk to Moscow of an influential member of the Scientific Council for DSc theses, Academician "Chernyaev" (as I will call him here), who could have had reservations about my work. They also advised me to choose some very strong opponents in case the move should not occur.

I still do not know for sure whether or not "Chernyaev" was as much biased or if he did not favor "item-five invalids," but I was greatly concerned anyway. However, Fradkin said at once:

"Let's ask Berezin to be an opponent; no Chernyaevs will stand up against him."

I did not get the point: was the man joking? Even though I was personally familiar with Felix Aleksandrovich, our research directions had no points in common and he hardly knew what I was working on; besides, I was somehow afraid to appear before Berezin himself with my simple, as I used to believe, mathematical tools. Nevertheless, Fradkin swept aside all my arguments of that kind. As I now realize, he gave me a good start in estimating the situation, and I am greatly indebted to him for that. Apparently, he could see that after reading Berezin's books I was inclined to generating functionals, canonical transformations, coherent states, functional integrals, anticommuting variables, operator symbols, and similar techniques that filled up my thesis. And besides, he could see that Berezin might be pleased with that.

At that time I used to think, incorrectly, that everybody knew those things and used them widely. After a while, Fradkin said that Berezin was waiting for me to call him up in order to discuss my thesis. I made the call, and he arranged a meeting at his apartment on Vinnitskaya Street. He met me at the anteroom, wearing sporting

Supreme Dissertation Council (VAC) for further consideration. VAC had the ultimate say: it could either confirm the degree sought, or veto the positive decisions of the local Scientific Councils. –Editor's note clothes and barefoot; he told me that was how he walked around the house. I did not pay much attention to all that, as I was nervously awaiting the discussion of my thesis. He said that he could take the final decision on being my opponent only after a detailed discussion of my work. Immediately after that we sat down to look through my thesis. He was calm and friendly, and he grasped the meaning at once. At the first meeting, we only had time to cover a small part. The next two meetings took place at the apartment where he lived with Elena Grigorievna Karpel, but I did not see her then (we got to know each other only years afterwards, as she writes in her reminiscences); their little daughter, Natasha, would run into the room, come up to the desk, and watch what we were doing. Berezin treated his daughter calmly; he did not use baby-talk with her, and he looked at her with a thoughtful expression. Finally, as we reached the end of my thesis he gave his consent to be my opponent, and I was sure that, at any rate, he had understood my thesis entirely, in contrast with many physicists (of course, one should not include Tyutin in the list of those), with whom I had discussed my problems for a long time already. It was also my impression that he had got involved in that business not only because Fradkin had asked him to, but also because he wanted to benefit from it and to find out what the others were doing.

Kolyma Stories

There were times during our meetings related to my thesis when we discussed some non-physical problems; he would ask me about my life, how I had come to live in Siberia, what was my opinion about this or that; he would make some remarks, and I was glad to know we looked at many things the same way. I am not afraid to say that we were on the same side of the barricade, which was a nontrivial fact, strange as it seems. Many things that appeared evident to me, the hatred for Stalinism and anti-Semitism, and for the system that was governed by yokels and arch-cynics of all kinds, the sympathy for Sakharov and Solzhenitsyn (of that time) and so on, were not the same even for highly intellectual professionals of physics and mathematics.

One day I told him that I had been twice, in 1973 and 1974, to Magadan, Kolyma. He asked me to tell him more about it. This was how it happened. At that time, being a young assistant professor at Tomsk Institute of Radio Electronics, I was assigned to be the president of the so-called "Outreach Commission for Entrance Examinations." During the summer vacations, the commission would depart to the "famous" — or, should I say "infamous"? — city of Magadan, the capital of the Kolyma lands, and organize entrance examinations to our institute for local youth.

Those two trips, each a month and a half long and full of active interaction with various groups of Magadan population, have made an indelible impression in my memory and transformed my previously bookish and schematic perception of the horrors of Stalin's system into a concrete one, saturated with life stories of ex-prisoners of Kolyma concentration camps and with images of those dreadful places (even in summertime). Later on, as I was reading *The Gulag Archipelago* by Solzhenitsyn, I told my friends:

"I think he didn't see the worst,"

though, obviously, I could see and learn during those trips only a small part of the things that had happened there. I recounted to Felix Aleksandrovich some of my impressions and told him a few stories.

One of these was how, upon our arrival by plane in the middle of summer, we went to the city from the airport along a road whose sides were clad in ice, and the driver was telling us about the Valley of Death, which either stretched along the road or was located someplace nearby. So, at any place of that valley the layer of human bones underground was several meters deep... .

Another story was as follows. As very important persons, we were taken by some military officer out to a usual entertaining voyage on a small military ship in the Sea of Okhotsk. As I have said, it was summertime but my raincoat was no use against the bone-chilling cold. The clouds, black and ragged, literally descended from the sky and touched the ship, with neither a bird nor any other living thing around us. We approached the coast at a distance of approximately half a kilometer. The coastline was a sheer wall of cliffs, steep, wild and barren. It was then that I realized that if hell actually did exist then it had to look exactly like that. And this is how it remains in my imagination until today.

Then, suddenly, out of the misty nothingness there emerged a frail boat, with some creature inside, and came near our ship. The boat was already expected there. In fact, as it turned out afterwards, its coming was a usual part of the entertainment program for the very important persons. It all happened in silence. Someone aboard the ship dropped into the boat a sack and a rope. The creature in the boat tied the rope to a huge basket, which was then lifted up aboard the ship. Then the boat disappeared in the mist. The basket turned out to be cram-full of enormous live crabs, intended to be cooked for a banquet aboard the ship. The man of the boat was said to be mute and legless, an ex-prisoner, without a name, without documents (I saw a lot of such people in Magadan in those days); he would come to that place to exchange crabs for salt and matches. Thus informed, I no longer felt like eating crabs and having fun in general.

One more story reached me like this. One of my Magadan friends, a journalist, took me to a sporting complex and introduced me to a person who worked there, a middle-aged man of athletic appearance; my friend had told me:

"The fellow is touched in the head, but they let him work here; he's quiet. Just ask him, and he'll tell you his life story; it sounds incredible even by the local standards."

As we became acquainted, I asked him how he managed to look so athletic at his age, and he said he could take me for a ride on a motor boat out in the sea (of Okhotsk) ... . In those days I was self-confident in every sense of that word, quite physically fit, and free of any fears and doubts, so I accepted this proposal with enthusiasm. As I now realize, my attitude had won him over, and as we went to the seaside he told me his life story. His family had been convicted in 1934, his parents and three children, he being the youngest, two years old, and they were on their way from the "mainland"[n] (apparently

[n]The population of Kolyma refers to the rest of the country as the "mainland." Nakhodka is a seaport in the Russian Far East. –Translator's note

from Nakhodka) in the hold of a ship bound to Kolyma. That was the very beginning (!) of convict transportation to Kolyma, which at that time was an absolutely wild and uninhabited territory (an area equal to France or something like that) covered by ancient forest. (They say that twenty years later, in the 1950s, any piece of timber had to be brought from the "mainland," since twenty years of "corrective labor" of the *gulag* prisoners completely eradicated these forests.)

Until now, communication with the remaining part of Russia is maintained only by sea or by air, and this circumstance has created an insular way of life, with the rest of Russia being considered as the mainland. Their ship came to anchor and the convicts were taken ashore by boat and then left there on the wild beach, alone and without any means of survival whatsoever (without food, clothes, tents, and even unguarded), to pass the winter until the arrival of the next convoy the following year. There were neither camps nor anything there at that time. The *gulag* had just taken root in that place; the territory was to be settled, and millions of convicts were to be put six feet under. During the first months the rest of his family and most of the convicts died from the cold and hunger, but he survived and turned into a wild beast. Then he spent decades in prison camps and in exile, and now there he was, a free man, who had a job but no family.

At that point we came to the beach, where there was a small aluminum boat (the kind known at that time as *kazanka*) with an outboard motor. Such a boat is unstable even in a quiet river, so taking it out to the open Sea of Okhotsk, which is typically stormy was nothing short of a suicide. Nothing could have forced me to get into that boat now, but then I had been impressed by the story and felt an unbearable pity for the man, who was actually long dead, so I got into his boat, and out we sped to the open sea. We rushed on through the mist, and again there was the ice-cold bone-chilling wind, and the boat was jumping from one wave to another, just as I was beginning to realize that in case it suddenly toppled over, the ice-cold water of the Sea of Okhotsk (in summertime, perhaps, +4 degrees Centigrade) meant certain death within a few minutes. The man was aft, getting the most out of the engine. He did not mean to scare or terrify me; he must have forgotten all about me. He seemed entirely immersed in his own story, once again attacked by the visions of his dreadful life. Apparently he was not clinging to life, and, more than that, was subconsciously hurrying the end. Nevertheless, we did get back. I will not forget the man and his story as long as I live.

As I remember, in connection with the Kolyma stories, I also related to Felix Aleksandrovich some confessions of a former *chekist* who had worked at the Far East in the GPU system and, after retirement, was the Head of the Personnel Department at our institute. He was on good terms with me, and he always helped me out and backed up my group. One day, being tipsy after a public demonstration, he opened his heart to me. I do not remember in which connection, but as I was under the impression of *The Gulag Archipelago* that I had just finished, I started talking about the horrors of convict transportation in so-called "Stolypin cars."° That caused him to remark,

"Do you know that prisoners sometimes managed to escape from trains on their way to the Far East? And some of them would be missed during roll-calls at railway stations."

He fell silent, and then he said:

"But shortage was not permitted ... Then the problem was solved like this: the guards would go to the nearest inhabited locality, grab the first people of appropriate age and appearance they could find there, and put them on the train to replace the runaways... ."

Of course, since Felix Aleksandrovich lived in Moscow, he was surrounded by a far greater circle of intellectuals critically disposed to the regime than the circle that surrounded me at that time, so in principle he knew everything about the system, but it appeared to me that my stories from Kolyma did not leave him unmoved. He told me once,

"Why don't you write about it, Dima?.."

Who could ever tell that his destiny was to die shortly afterwards, and exactly there, in Kolyma, and that I would include these stories in a collection of reminiscences about him?

°Cattle wagons barely adjusted for the transportation of prisoners. –Translator's note

Novosibirsk 1979, Akademgorodok, Thesis Defense

So, Berezin was my opponent and my defense was to take place in December 1979 in Akademgorodok,[p] at the Institute of Nuclear Physics, Novosibirsk. Accompanied by a support team (Tyutin, Bagrov, Lavrov and Shvartsman), I arrived in Akademgorodok by plane in the morning one day before my defense. The frost was unusually severe, even for that region, about 45 degrees Centigrade below zero. Berezin was to arrive in the evening. I was allotted a GAZ-69[q] and drove out to collect him at the airport, having taken along, just in case, somebody's fur hat with earflaps.

That hat came in very handy. An incredible thing had happened. The plane had arrived an hour ahead of schedule, and when my truck entered the square in front of the airport terminal, there was Berezin standing there in the frosty evening gloom, stiffened in his light overcoat and without a hat. Those who have ever experienced minus 45 will appreciate my horror (Tyutin will not). However, he did not let on that he felt cold or that something was wrong. He put on the hat, though, and we rushed on to the hotel, where the support team, with yet another opponent, Barbashov, had already laid the table, serving it with a generous quantity of vodka.

I could not imagine that I would ever drink the night before my defense; however, all present, including Felix Aleksandrovich, came up with so many arguments in favor of drinking that I gave up on that and the fun began. I was eating and drinking with the others, feeling all along a bit frightened in view of the coming event; at the same time, one part of me was intent on watching Berezin in such an unusual setting.

Of all things that happened that night, one thing that I remember most, and will not forget as long as I live, was the behavior (I am not sure that the word "behavior" is the right one; perhaps I should

[p]Akademgorodok (Academia Town), a town located 20 km south of Novosibirsk and founded in the 1950s by the Academy of Sciences of the USSR to be one of the educational and scientific centers of Siberia; at its peak, Akademgorodok was home to 65,000 scientists and their families, and was a privileged area. –Translator's note

[q]A four-wheel-drive light truck produced in the Soviet Union between 1953 and 1972. An analog of Jeep. –Translator's note

say "reactions") of Berezin. I will attempt to describe what I saw and to give my interpretation of it. The mission of those present at the table, as they understood it, was to cheer me up, inspire me with self-confidence, and even insolence, in view of the coming defense. They rudely jested, often using obscenities, telling indecent jokes, laughing out loud, etc. If it had not been for Berezin, I would consider it normal, but that time I thought that the atmosphere did not chime in with his person and his presence. As I wanted to cool the fellows down, I watched the reaction of Berezin. It was so touching to observe: he definitely did not belong there; it grated on him to be exposed to ribaldry, yet he made his best to conceal it, as he did not want to be looked on as a *rara-avis*; he desperately, perhaps just once in his life, wanted to be like everybody else. He laughed at indecent jokes and joked himself, but he was incapable of any vulgarity. He certainly was a *rara-avis*, even between us, not to mention the average population of Russia (I am perfectly familiar with that sensation of a *rara-avis*; in Siberia I was a stranger, a Jew, who looked distinctly different from Siberians; in Moscow I was a provincial from Siberia and a Jew to boot; in Brazil I am an aggressive Russian; even in Israel, where I came right after the beginning of Perestroika, thinking they would receive me with outstretched arms, I was told:

"You come here and you are all provided for, and we had to..." and so on).

The next day we all went to my defense at the Institute of Nuclear Physics. That day, the Council had two DSc theses on its agenda, the first one was Max Zolotorev's (Budker's son-in-law and the closest collaborator of Barkov, the President of the Scientific Council, with whom Zolotorev discovered parity nonconservation in atoms), and the other was mine, of an unknown Gitman from Tomsk. While Zolotorev was performing his role, all calm and confidence, I was feeling quite depressed. A piece of Jewish Luck: in the morning the news had spread around that the day before that Akademgorodok had been blessed with an illustrious visit of Aleksandrov, President of the Academy of Sciences of the USSR, who had arrived from Moscow, with a retinue including Academician "Chernyaev" whom we had already written off (with great joy) as a no-show. More than that,

"Chernyaev," as he remained a member of the Scientific Council, had promised to seize the opportunity and visit the Council's meeting.

Indeed, I had hardly begun my report when "Chernyaev" came in the hall; in excellent spirits he seated himself in the first row and took my thesis from the secretary. Then there comes the following pantomime: "Chernyaev" opens up the first page; his smile slowly fades away from his face; he turns around to the audience, more precisely, to that part of it which contains most of the Council's members, his face wearing an expression that reads like this:

"Just what do you scoundrels think you are doing here in my absence?"

He then assumes his initial attitude and starts to leaf through my thesis with a look of disgust. I realize that my number's up and just keep on mumbling through my report, expecting the worst. Now, the report is over, and it is time for questions. "Chernyaev" puts a question:

"What is the practical importance of your work?"

I mumble out something in response; indeed, what can be the practical importance of the quantum creation of particles in supercritical fields, which actually do not exist in Nature? I feel nasty, looking for salvation in the profound thought

"To hell with all that."

I return to my seat and wait for the shameful end. And then the floor is taken by Opponent Berezin. He strides resolutely to the rostrum and, addressing "Chernyaev" personally, begins his speech, to quote (I have its tape record):

"Esteemed Professor, we all know how great Lev Davydovich Landau was; nevertheless, he also made mistakes. For one, he never accepted Feynman diagrams. But that's understandable, because Landau was able to calculate any process without the help of diagrams. But he was Landau, after all, whereas for common scientists those diagrams have always been very useful. I think you are running the risk to repeat that instructive mistake of Lev Davydovich and underestimate the diagram technique developed in Gitman's work. You know, it may prove very useful in the solution of a number of problems... ."

It was no longer important what was said after that. One could see that "Chernyaev" was stunned at first; then he thought for a while and then started nodding his head as the comparison with Landau had its effect on him; Berezin had won the game. At that moment I merely was happy; Berezin had saved me and showed everybody that he had really understood and appreciated my work. Only now, after reading the reminiscences of Elena Karpel, do I know that his speech at my defense was not just a lucky improvisation; in fact, he had prepared himself very seriously for any possible confrontation, and he was pleased to find a brilliant way to disarm and defeat his opponent; it is this brilliant talent of Berezin as a polemist that is mentioned in the reminiscences of Elena Grigorievna. One could see that the mood of the audience had changed; even "Chernyaev" voted in my favor.

In the evening, there was a banquet for a small company, my friends that had arrived with me, Chaplik, a friend of mine who lived in Novosibirsk, and my opponents. I was so happy and grateful to everybody that in the middle of the banquet I asked all present for a permission to propose a toast to each one in turn. I started each toast with some characteristics of the person in question (of course, it was not modest on my part, but I was carried away with emotions; I wanted to say something good about everyone, and I did have what to say).

Those present at the table had different scientific weights, both formally and actually, but it was clear that the best-known person there was Berezin. In spite of that, one could see that he was pleased to listen to good things addressed to the others, and it never occurred to him that he was the star of the first magnitude among those present, and the good things said to the others did not affect him in any way. Of course, I said many good things about him as well, and about the way he had helped me out, and then also about the others present at the table. Then, when it was time to propose a toast to Tyutin (who was much younger than Berezin and was not well-known at that time), I said just one phrase, which I had thought I should never say in the presence of Berezin, but I did say it, anyway:

"And now I wish to drink to the health of a person who knows everything in theoretical physics."

Of course, the professionals there at the table appreciated the code transferred by this phrase rather than its absolute meaning. And then there happened a thing that I will never forget, a thing that completed, in my perception, the impeccable image of Felix Aleksandrovich as a person and scientist. He stood up abruptly, with a wineglass in his hand, as though he had been waiting for it all the time, and, with a gladness which I understood, exclaimed:

"Yes, let us drink the health of a person who does know it all, and from whom I have learned a lot of things."

That night I saw Berezin for the last time.

1980 and Later

During that banquet we had arranged with Felix Aleksandrovich that he would come to Tomsk in the summer of the following year 1980 to present a course of lectures and to participate in a short hiking tour through the taiga. In the beginning of the summer of 1980, we got in touch over the telephone a few times, discussing the details of his visit. However, he suddenly called me up in Tomsk and excused himself, saying that a friend of his had arranged for him an opportunity to travel through Kolyma ...

He did not mention any details. I had an eerie feeling. The point was, that friend of his, a remarkable person and scientist, had quite a dubious reputation in tourists' circles. He was a fan of hiking tours, and it was known that in the tours he participated something always went wrong. Of course, he did not have anything to do with this, but the black-cat effect did happen, anyway. I was really sorry. I wished Felix Aleksandrovich to have a pleasant tour, and we arranged to postpone his visit to Tomsk until the following summer. I promptly changed my plans for the summer; however, in July I was still in Tomsk. It was there, in the middle of July, that I received a phone call from Tyutin who told me that Berezin was no more. My world, which mainly consisted of a dark background with only a few bright spots, suddenly turned darker. Nobody knew all details of the tragedy, and I, being prone to pessimistic outlooks, and, apart from that, being very angry at the system by that time, imagined the worst. Now, after reading the reminiscences of Elena Karpel, I can see that, unfortunately, I did have some reasons for that... .

Time passed; Tyutin and I were engaged in quantization problems, which had been so interesting to Felix Aleksandrovich; we always applied his ideas and methods, and even wrote a book on that subject,[r] which we, naturally, dedicated to him. The first page of this book reads:

"*We dedicate our book to the memory of the prematurely deceased outstanding scientist, mathematician Felix Aleksandrovich Berezin, whose influence can hardly be overestimated. Felix Aleksandrovich took great interest in quantization problems and, to a great extent, stimulated us in our studies.*"

At the end of the 1980s, we rediscovered for ourselves one remarkable work of Felix Aleksandrovich, written in 1975–1977 together with Misha Marinov and dedicated to the quantization of a spinning particle.[s] This work, in fact, reveals a different side of Grassmann variables. They not only ensured a correspondence principle for Fermi fields and particles, but also permitted, as a result of quantization, to obtain quantum spinning systems with finite-dimensional Hilbert spaces. Later on, I spent a lot of time working with different modifications[t] of that first model of Berezin–Marinov, automatically confirming my fatal devotion to his ideas and methods. I still nurture some plans in that direction.

The reader already knows from the reminiscences of Elena Grigorievna about my acquaintance with her (not just a coincidence, I suppose) in the late 1980s. This acquaintance substantially enriched my perception of Felix Aleksandrovich. I think it would be fairly impossible to imagine another woman by his side. And their daughter,

---

[r]D.M. Gitman and I.V. Tyutin, *Quantization of Fields with Constraints* (Springer-Verlag, Berlin 1990).

[s]F. A. Berezin and M. S. Marinov, *Particle Spin Dynamics As The Grassmann Variant Of Classical Mechanics,* Annals of Phys. **104**, 336 (1977).

[t]D.M. Gitman, *Pseudoclassical Theory of Relativistic Spinning Particle*, in *Topics in Statistical and Theoretical Physics*, Berezin Memorial Volume, American Mathematical Society Translation Series 2, (Providence, RI, 1996) vol. 177, pp. 83-104; D.M. Gitman, *Path integrals and pseudoclassical description for spinning particles in arbitrary dimensions*, Nucl. Phys. **B488**, 490 (1997).

Natasha, who at that time attracted my attention by her inner and outward resemblance to Felix Aleksandrovich... I have never met her since then. But it seems to me I will meet them again.

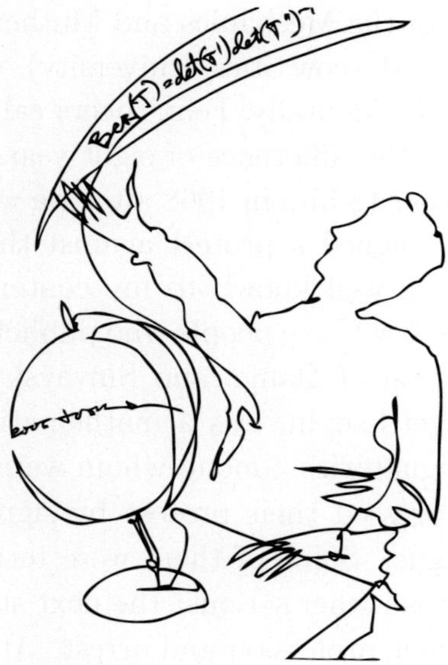

ABOUT ALIK BEREZIN AND HIS TIME *

VICTOR PALAMODOV
palamodo@post.tau.ac.il

Around 1964, soon after finishing graduate school, I found myself in a subdivision of the Mechanics and Mathematics Department (mekh-mat) of MGU (Moscow State University), where Felix Berezin worked at the time. Incidentally, Felix' peers called him Alik, but I was much younger — the difference of eight years seemed greater at that time. I grew closer to him in 1968 after we were amongst the 99 mathematicians who signed a protest against the arrest of Esenin-Volpin, whose story is well known to my contemporaries. Esenin-Volpin was one of the few brave people who publicly protested against the judicial persecution of Daniel and Sinyavsky.[a] Esenin-Volpin, son of the poet Sergei Esenin, was a mathematician by profession. Berezin and other signatories, among whom were rather well known mathematicians, expressed their protest by signing a letter in defense of their colleague. Some of them were terminated from their employment. This was rather serious: the next step could have been being barred from their profession and arrest. At the same time, for the faculty members of the mekh-mat Department no serious punishments ensued. Later, I began to ponder this. Below I will attempt to provide an explanation.

Alik's road in life was a harsh one — fatherless and itinerant during the hungry post-war years, he finished mekh-mat in the memo-

* Translated from Russian by Roman K. Kovalev, The College of New Jersey, Department of History, Ewing, NJ 08628, USA; e-mail: kovalev@tcnj.edu.

[a] Yuli Daniel and Andrei Sinyavsky were Russian writers. In 1965, Daniel and Sinyavsky were arrested and tried for their dissident activities in the infamous Sinyavsky–Daniel show trial. On February 14, 1966, Daniel was sentenced to five years of hard labor in the *Gulag* camps for anti-Soviet activity and Sinyavsky to seven years. Unprecedented in the USSR, both writers plead not guilty. – Editor's note

rable year of 1952. This was the peak of the disgusting Judophobian campaign usually referred to as "The Case of the Doctors." [b]

Alik, along with Robert Minlos, was recommended for graduate studies at mekh-mat, but was barred from the entrance examinations. Instead, he was sent for some time to Mongolia to teach mathematics in a local school. I remember a story he told about this time. Incidentally in Mongolia he did not feel humiliated while in MGU this feeling was omnipresent.

The death of Stalin and the subsequent political developments brought an end to his "punishment." He was allowed to pass the entrance examinations and was accepted to graduate school in 1956. Having quickly advanced up through two academic levels [c] — candidate and doctorate dissertations — and having shown a bright talent, Alik seemingly could have started thinking of the professorial position at the aforementioned division. From the point of view of science, Alik and Robert Minlos had no competitors. Neither of them ever received that position. Why?

There were no obvious objections to their work and to the behavior of the contenders! The question "why" appears naive. As I recall the logic of Soviet power there was a dire obstacle — a "wrong" nationality. [d] Being barred from graduate school in 1952 due to his "doctors'" nationality, [e] Alik received a mark of disloyalty in his dossier which could not be expunged because of such silly rea-

[b] At the end of 1952 a number of medical doctors, the most accomplished medical specialists, were arrested on fabricated charges. All of them, with the exception of two, were Jews. They were accused of acting under the orders of an American Jewish "spy" organization, *Joint*, with intention to kill the leaders of the Communist Party and the Soviet Government under the pretense of healing them. It was only Stalin's death that saved them from capital punishment. Shortly after Stalin's death it was announced that all charges were fabricated. – Editor's note

[c] The academic hierarchy in Russia follows the German rather than the Anglo-American pattern. An approximate equivalent of PhD in the US is the so called *candidate* degree. The highest academic degree, doctoral, is analogous to the German *Habilitation.* The doctoral dissertation is usually presented at a mature stage of the academic career; only a fraction of the *candidate* degree holders make it to the doctoral level. –Editor's note

[d] In Russian the word "nationality" is used to mean ethnic origin. What is known as nationality in the West is called citizenship. –Editor's note

[e] An allusion to The Case of the Doctors; F. Berezin was Jewish. –Editor's note

sons as complete acquittal of those doctors who gave the name to the infamous campaign. Simply put, newspapers and what you read there — this is for the commoners.

Incidentally, mekh-mat's Administration could have done things differently. Instead of admitting Alik to graduate school, they could have very easily sent him to Uriupinsk[f] (after finishing an institute of higher learning former students had to go through a process of "allocation." Administration had the right to choose for them the place of work for a mandatory three-year term). This did not happen in Alik's case, however. Luckily for Alik, his qualification and personal abilities apparently were taken into consideration.

Mekh-mat's bosses amassed at mekh-mat excellent researchers whose pedagogical and scientific potential greatly exceeded the demands of the department itself. Again, why?

In my memory, aside from the Mongolian story, there is another one. Several fellow students were sent to Guinea; several colleagues moved (or, better to say, were ordered to move) to Cuba. Someone was also sent to Afghanistan during the period of Soviet intervention. It is clear that we were viewed as a component of the third echelon of Soviet expansion around the whole world (first and second echelons are the military and political divisions).

Who personified power at mekh-mat? Limited functions of power, of course, belonged to the Dean, but the Party Committee could depose (or, should I say, gobble down?) him, as it did happen once. The composition of the Party Committee was no secret and was periodically renewed, but its policies changed little. The "guiding orders" came through the Party Committee of MGU that was under surveillance by the Central Committee of the Communist Party of the Soviet Union.

Obviously, there could be no illusions about traveling abroad. This prohibition was initially considered as one of the many prohibitions and constraints of the era: not to listen to "enemy" radio, not to read journals which were published abroad in Russian, and available, supposedly legally, to certain individuals, etc. Traveling abroad was

[f] Synonymous to "in the middle of nowhere." –Editor's note

conceivable (quite rarely, though) only for especially trusted people. For Alik and other active scientists, the ban on communications with foreign colleagues was a very sore issue, and he did not hide this in our conversations.

The feeling of hopelessness was terrible — even rays of hope that this and other prohibitions would be loosened seemed inconceivable. Soviet power seemed eternal in the scale of human life! In reality, the beginning of its collapse occurred then, in 1953, when new leaders refused to carry out mass purges and total repression — without these tools the system began to fall apart. Very few people understood it back then, when we were young. We worked with relish and thirsted for discourse and immediate recognition.

The problem of travel abroad and communication with foreigners was a challenge for the Soviet authorities at that relatively "humane" time. With the previous leader, Stalin, it was much simpler. Soldiers of the Red Army who met up and interacted with the Allies in 1945 on the Elbe were soon arrested and interned into camps. Now, after Stalin's death, everyone who had been abroad and did not defect presented a problem since his or her loyalty after becoming acquainted with the West could be put under question.

There was no official prohibition on legal crossings of the border. Such a prohibition would have complicated the position of left-wing parties in the West, which presented an important source of pressure on western governments with the aim of weakening their opposition to Soviet power in connection with its expansion (anti-war campaigns in the West were also encouraged as another useful tool of pressure on western governments). In this situation, the authorities chose palliative measures: formally not to refuse the right of the "population" to foreign travel, but to sabotage its realization in various ways.

Control over foreign correspondence entering MGU was one of such measures. As I was told, such correspondence was sent in bags to an undisclosed location in a building belonging to MGU and after being "worked over," which would normally take a week or more, appeared in large portions at our department. This was noticeable since all of the mailboxes were simultaneously filled with packs of letters in airmail envelopes of standard type. This "imperceptible" service, apparently, could not cope with the stream of letters, many of which contained invitations to scientific meetings. An entire drawer in Alik's desk was filled with such letters. His attempt to go to CERN (Geneva) for an extended period of time was met with a determined refusal. Thereafter, as he appealed to the new President of MGU, R. Khokhlov, with a letter explaining the situation at the department, Alik was refused all possibilities of traveling abroad.

I was in an analogous situation but from time to time I still made attempts at least to start the process of preparation of appropriate documents so as to test the resistance of the system. The procedure included receiving some twenty signatures or "visas" and going through several Party interviews. Usually, the affair came to a halt at the level of our department. Once, I recall, I brought a letter with an invitation to the head of our departmental "MID."[g] Having read the letter and badly concealing his irritation, he raised his colorless eyes at me: "Where is the envelope to this letter?" I showed him the envelope. Having examined it, he said: "You will have no time to go through the medical commission and finish the paperwork by the deadline."

This answer did not surprise me, but I became vexed over the question of why he looked over the envelope. Apparently, he was convinced that the envelope was not inspected. The primitive method of unsealing them by opening the envelope with steam would have left a noticeable trace, taking into account the type of paper it was made out of. Later, I was told that the method of unsealing envelopes at MGU was upgraded, apparently, to inserting a catheter with a miniature camera into the envelope by way of which the text could be brought up on a screen. Another method involved analyzing letters that were written on half-transparent paper by way of shining a very strong light; the image received from the light was processed with a special computer program which digitally "unwrapped" a page of the letter, which had been folded threefold in the envelope. I, however, do not believe that such a sophisticated method could work well with

[g]MID is the Russian abbreviation for the Ministry of International Affairs. The author means mekh-mat's Office of International Relations. –Translator's note

the university's KGB functionaries.[h] At a certain point, I began to notice that the envelopes that I received at the department lacked a side edge — it had been simply cut off with a sharp razor!

Ironically, Alik was not totally bereft of communications with foreigners. The Party Committee of the Department ordered him to become a scientific advisor of a student who came from Egypt. During one of our long walks in the forests outside of Moscow, he told me with some humor that this Egyptian student did not spend too much time at work. He understood Alik's situation very well — Alik could not get rid of him and was supposed to personally write a thesis for him, if necessary. Alik was personally responsible before the Party Committee in his defense. This continued until the President of Egypt G. Naser died. A. Sadat took office after him. Sadat's relations with the Soviet Union were quite different. The Egyptian graduate student quickly understood that people would no longer do work for him and began to study in earnest.

During those walks in the forest we, of course, could not avoid discussions of political issues (although they did not bring us any pleasure). Politics were everywhere. Once I showed Alik one of the maps of the Moscow region, the so-called "tourist map," naively pointing out the obvious inconsistencies of geographic places that were well known to me. Alik explained that this was not a misprint: errors and distortions were introduced on purpose!

"Why?"

"So that we, i.e. simple tourists, did not feel ourselves to be the owners of our own country!"

He showed me, at the corner of the map and in small print, a statement to the effect of "compiled by GUK" and said: "This is a division of the KGB."

Sometime during the 1960s, I became involved with the stream of

[h]KGB is the Russian-language abbreviation for State Security Committee, the principal secret police agency. The Soviet secret police changed acronyms many times. It started out as the Cheka, and then became the GPU, the OGPU, the NKVD, the MGB, and, finally – since 1954 till the demise of the Soviet Union in 1990 – the KGB. The term KGB is also used in a more general sense to refer to Soviet political police since its foundation as the Cheka in 1917. –Editor's note

literature from *samizdat.*[i] I recall that the first such text which I read was the well-known program of A. D. Sakharov. Alik, of course, had access to the samizdat through his own channels and, in general, was better informed. I found out from him that Sakharov was one of the creators of the Soviet hydrogen bomb. In that program Sakharov discussed capitalism and communism, their ability to coexist and their actual or expected convergence.

In all of my simplicity, I understood that Sakharov's ideas were naive; the sinister meaning of his ideas came to me much later. This was the beginning of the evolution of A. D. from being a genius weapon-maker of the regime to being a human rights activist. This image overshadowed the previous one.

Sakharov's role in our recent history is contradictory. He, together with his colleagues, gave to the cruel regime the technology of the hydrogen bomb. It can be recalled that the first testing of this weapon occurred in 1956, and soon the Red Army intervened in Hungary. The interconnectedness of these events is obvious. The death of the Soviet system without the hydrogen bomb could have been much quicker, and Alik could have seen the "light at the end of the tunnel."

Only much later did I understand the meaning of Sakharov's prophesy. These systems — capitalism and socialism — indeed, do converge in some way! The concentration of power in the European Union growing out of the Brussels bureaucracy, pseudo-ideology in the form of political correctness and exultation of sodomy, hypocrisy and double standards, and, finally, armed intervention in South-Eastern Europe! I think, however, that this analogy is not complete: Europe will not repeat the development of the USSR in detail; rather

---

[i]A strict censorship existed in the USSR. Nothing could be published without pre-approval from *Glavlit*, an omnipotent State Agency implementing censorship. The class of suppressed books and other printed materials included not only those with political connotations, but, in general, everything that was not considered helpful for Soviet ideology. Forbidden publications circulated in typewritten form. People retyped them, using mechanical type-writers and carbon paper, or photographed them, page by page, using amateur cameras, and then printed them at home on photopaper, producing huge piles. The process was called *samizdat*, which can be loosely translated from Russian as self-publishing. *Samizdat* was forbidden by the Soviet law. – Editor's note

Europe is moving towards a fall along its own trajectory showing similar symptoms.

Thus, travel abroad was impossible for Alik; communications with foreign colleagues were rare and restricted (KGB closely shepherded the participants of the International Congress in Moscow in 1966). The space of scientific activity inside "the large zone, one sixth of the world" (i.e. USSR), is limited due to the severe scientific and personal climate. Independence in behavior and especially in his scientific position was the key element of Alik's character. Under the conditions of semi-feudal relations in science, any academic career was significantly complicated for those who did not belong to an entourage of some scientific "warlord." I recall a question by one provincial petty mathematical warlord addressed to me:

"And who do you work with, Victor Pavlovich?" i.e. for whom do you work?

It was impossible to imagine Alik working for someone or choosing a field of research just because of some authority. I. M. Gelfand, active at that time in the department, initially helped Berezin when he was a student and then in graduate school. After a bit of time, Gelfand began to consider Alik as his competitor, who early on became an independent researcher. A serious conflict ensued, the meaning of which Alik explained to me, but I will not describe it. As a result, Alik ceased to attend Gelfand's seminar, and, in general, he cut off all contacts with I. M.

Furthermore, Alik warned me and quite likely his other colleagues that he very much did not want to have Gelfand among the signatories of Berezin's obituary! Such a situation — Gelfand signing Berezin's obituary — at that time seemed remote and highly unlikely. But it did happen and, unfortunately, Alik's decisive request was ignored.

Physics, more concretely new mathematical physics, was Alik's primary field. Apparently, the greatest influence on Alik was the famous seminar of Lev Davidovich Landau. It is known that active participation in this seminar was not for those who "had no guts." Once Alik was so shocked with Landau's sharp remark towards him that he told me an aphorism which I remember: "The caliber of a physicist is characterized by the amount of harm which he can bring to science." The severity in relationships, of course, was caused and intensified by the inhumanity of the regime and the stern constraints imposed upon us all. At the same time, different people behaved differently.

In these conditions Alik was still able "to sing his song." Robert Minlos wrote in detail about the significance of Berezin's contribution to science. In my memory, Alik was completely preoccupied with creating supermathematics as this field was then called. He sought out superanalogs for basic algebraic and analytical constructions and formulated them in a systematic way. But he had little time remaining for this ...

Alik's strong impact on noncommutative analysis is undeniable — the term *Berezinian* entered into mathematical physics for good. The general conception of quantization gained more recognition over the years. The contemporary development in this field, in which Alik's ideas played a significant role, is remarkable compared to the level of the 1970s. Alik wrote several important books; some of them have been translated into English. Some Alik's books were published after his death, in the 1980s and '90s.

Not everything in our lives was gray and dim. An opportunity to find oneself far from the city, for example in the forest, was the best medicine for Alik's psyche, overloaded with negative emotions. Yes, in the forest or elsewhere, where there were no buildings or ruins reminding us of the omnipresent regime. Hiking, long walks, bicycle and ski trips were his main pleasures and methods of preserving spiritual balance. There was one more important motive in his choice of paths for his wanderings...

In the summer of 1969, a group of more than 50 mathematicians, Alik amongst them, got involved in an escapade that was organized by the Krasnoyarsk mathematicians. A ship with the participants of the scientific conference onboard embarked from Krasnoyarsk down the Yenisey River in Siberia. After several days of travel, we found ourselves amongst deserted banks covered with dense taiga. It was hot June weather, but we were unapproachable for the mosquitoes being in the middle of the huge river. And on the 20th of June, our vessel crossed the Arctic Circle not far from Igarka. We observed the sun slipping above the horizon several days in a row. On our stops we jumped from the deck into the warm Yenisey water, dark from the rotting timbers which lay at its bottom. From Dudinka we journeyed along the railroad to the town of Norilsk, which has special significance for Soviet history. Alik carefully peered into the ruins of the *Gulag* camps that appeared along the way. Remains of barbed wire and leaning poles were visible along both sides of the road. This was not a festive curiosity. He undoubtedly attempted to imagine the *Gulag* life, which was known to us thanks to *samizdat*, and himself in this life.

After our ship returned to Krasnoyarsk, a group of fifteen people, including Alik, Bob Minlos, and Sasha Kirillov went off on our next adventure. We went hiking across the taiga by the river Mana. In the stuffiness of the moist forest, the insects finally had their chance. On the advice of Alik, the voyagers, all men, took preventive measures. We undressed to the buff, leaving on only footwear and loaded all the clothing into our backpacks. Moving in a line along the dense taiga, each of us regularly looked over the back of the one walking in front, seeking for ticks which jumped from the branches, orienting by smell and climbing along the backs as they attempted to find a convenient place for a suck. Other insects did not disturb anybody, as they were not the carriers of brucellosis and encephalitis. Coming out towards the river after several dozens of kilometers above the confluence into the Yenisey, we got ready for rafting. Alik, taking on the role of the leader, began confidently fastening the logs. The timbers that were prepared up the river were simply tossed into the river and allowed to drift. Freshly cut timbers lay on the banks at any place along the river. The raft turned out to be excellent and we, in wild enthusiasm, came through all rapids passing other rafts that were built with less qualified hands.

Alik's way of life was formed during a difficult childhood and fatherless years (as among many other of his contemporaries). In his appearance and clothing there were no signs of "officialdom" such as ties, white shirts, etc. He did not wear anything like blazers, apparently so as not to be in any way seen as the people surrounding him who were functionaries and conformists; only a checkered shirt of bleak colors with the collar never buttoned down, and that was all. He put on his glasses only in cases when he absolutely had to; apparently, he considered that they did not correspond to his image. An old-style briefcase completed his ascetic style which for him was organic. It was clear that at home no one put pressure on him with the aim of giving him more of a conformist appearance. One could apply to him the rhetorical phrase of Ostap Bender[j] "Do I seem to be a person who has relatives?"

This phrase in an even greater degree characterizes his mother, Esfir Abramovna, his only actual relative.

In this same way, his ritual of socializing with friends and acquaintances was rather simple. Once he began a discussion with me on an unusual topic; he told me with soft humor about one unknown-to-me woman with whom he apparently became recently acquainted. In his words, I felt an unusual melody. A year or so passing by, I appeared at his house to pick up some text, and I discovered that his apartment was totally cluttered with children's diapers. But my visit did not become a reason for formal acquaintance with his young wife and Natasha, his newborn daughter. New circumstances did not change Alik's lifestyle. I met his wife Elena Karpel and Natasha much later ...

In the summer of 1980, Alik got the opportunity to travel to Kolyma as a member of a geological party. This was not his first attempt. Much earlier he had already tried to get a chance to achieve his dream: to visit THAT place. In 1980 the dream came true. The word "Kolyma" says a lot to our contemporaries, as it is described in the memoirs of those who lived in the countless *Gulag* camps. Kolyma was understood as a system of forced labor camps located in the basin of the river under this name. Due to geography, it is apparent that the climate is extremely severe and escape is impossible. The geological party in which Alik worked was supposed to return by way of rafting along the river Seimchan, a tributary to the

[j] An intelligent con artist, the main character of the satirical dilogy *The Golden Calf* and *Twelve Chairs* written by Ilya Ilf and Evgeny Petrov in the 1930's which was very popular in the Soviet Union. –Editor's note

Kolyma. I was greatly jealous of Alik, but something happened that was not at all expected. I did, though, have a chance to see Magadan, Kolyma's capital. I was returning to Moscow together with Lena Karpel, accompanying Alik's coffin. The bitterness of the loss filled my soul.

Science for Alik was a method to overcome the barrenness of the physical space-time to which he was confined. His science remains with us and with the rest of the world forever. I see in Alik's daughter reflections of several important features of his personality.

These hard times, as it seems, are over. But the future is indeterminable and alarming, and there appear to be no serious factors which would preempt new difficult times.

Ashdod, Israel
May 31, 2006

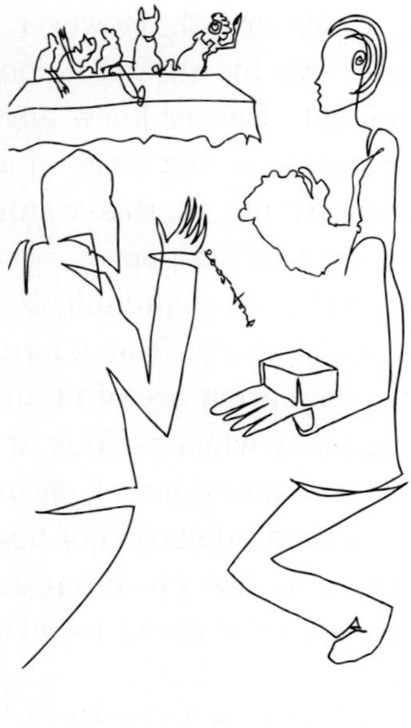

WITH AND WITHOUT FELIX BEREZIN

DIMITRY LEITES
Max-Planck-Institute for Mathematics in the Sciences
Inselstr. 22, DE-04103 Leipzig, Germany[*]
leites@mis.mpg.de

With him it was often difficult but fun, without him it is bleak...

Introduction

"You can not explain the bloom, the charm, the smile of life, that which rains sunshine into our hearts, which tells us we are wise to hope."

Anna Strunsky,
from a letter to Jack London.

Black haired (where not grey) and with shining blue eyes, Berezin was quite distinctive in any crowd. However, what struck me most from the very first sight, was his aura of a doomed one, as if a pending tragedy were imminent. I never knew anybody who perpetually looked (to me) so alarmingly at the verge of a disaster.

I was hesitant to contribute to this volume and stalled till the latest moment. There are two reasons:

1) It is next to impossible to communicate the "atmosphere" and tacit rules of a society to the reader from a foreign society: The "facts of life" learned by one generation are of no interest to the next one, as is, unfortunately, manifest when we look at the repeated mistakes being performed by the populations of entire countries again and again. The very few who read books do not make politics, and among those who do read, precious few are interested in mathematicians. Even in, somewhat spicy, stories about beautiful minds of "prophets

[*]On leave from Department of Mathematics, Stockholm University, SE-106 91 Stockholm, Sweden; e-mail: dleites@math.su.se.

without honor in their own country." To get a glimpse of the "feeling of place" at that time, I advise the reader to begin with the stories by V. Shalamov and Yu. Dombrovsky. Otherwise any narrative about life in the Soviet Union should be punctuated, like by hiccups, with clarifying but still poorly comprehensible footnotes.[a] For that reason — to convey the mood of the times — the anecdote in what follows is prefaced with several, far too long, paragraphs about myself, rather incongruous in the volume about Berezin.

2) How to impart what looked to me as Berezin's unceasing pathetic struggle against destiny and, sometimes, as attempts to escape from the burdens and various moral obligations in which he shackled himself, by inviting an ultimate *force majeure* to his rescue?

But I owe this effort to Berezin: He raised what instantly looked to me as an immensely tantalizing problem whose solving (constructing supersymmetry theory) occupied all my life since then.

Despite his occasional attempts to look rather distant and cold, Berezin could not deceive anybody: He was as protective of his students as the hen mother is of her chickens. He refused to be my official scientific advisor, and asked Arkady Onishchik (also not a "teacher's pet" in the eyes of the authorities) to take me, which Onishchik did (for which I am thankful to both of them and to fate) because, Berezin explained to me, "the authorities will not let my student enter the Ph.D. school." I was very offended. My hard feelings dissolved several years later when I was denied the right to enter Ph.D. school, because, I was unofficially told, "you are a student of Berezin." Although officially I was affiliated not to Berezin's Chair and Onishchik's recommendation was supported by Manin and Vin-

[a]Like this: For example, Berezin's letter to Moscow Mathematical Society was triggered by defenses of Doctoral (sic!), not Candidate's, theses with (according to experts) no (or dubious) scientific results whereas no scientific council agreed to accept for defense a result rewarded with the Fields medal (this other story began somewhat later and lasted for almost a decade, until the mathematical bosses died and Gorbachev's *perestroyka* started, but similar things kept happening meanwhile). The reaction of my other teacher, J. Bernstein, to this was remarkable and noteworthy. During a discussion of his colleagues in the kitchen (the then habitual analog of the living room) of his apartment "why doesn't he (the Fields medalist) fight? he should!" JB, usually silent, wedged in: "Why should he? He is a mathematician, not a wrestler!"

berg.

Berezin talked to me more openly than to his other students. In particular, he told me a bit about himself. For example, he told me how he was given his first job.[b] This was during the time the notorious "The Case of the Doctors" [c] was in full swing. Berezin was summoned and magnanimously told, "Your letter of recommendation from your scientific advisor [presumably, I.M. Gelfand] states that you are talented in mathematics and capable in research. So we would like to give you a chance to work at equal proximity to THREE world famous research centers: Odessa, Lvov and Moscow."

This paradise on Earth was a village in the middle of nowhere.

However, several days after this, Stalin died, and about a month after that, most such nominations were reconsidered (though not without efforts on the student's part). Berezin, rather than being recommended to the University's Ph.D. school, was given a position of a teacher at a middle school, but still in Moscow. (Many bright *and* independent students had similar fate, and not only during the grim times of Stalin. The recommendation for Vera Serganova, one of my brightest pupils, officially affiliated to Yuri Manin, was turned down by the Party group of the Chair of Algebra in the mid 1980s. Same thing happened in 1975 to Irina Shchepochkina, a student of Kirillov. (These two examples are the ones of the brightest and closest to me among many that took place.) And these girls were not Jewish at all, as a hasty reader might have surmised.

---

[b] At that time, everybody who was not pregnant or in some other way obviously exempt, was assigned for two or three years to a compulsory job upon graduation. Representatives from factories, research institutes, various Ministries, and so on met with students during a week prior to the days on which the students defended their M.S. diploma theses and each side tried to make their pick. The talented ones (and not so talented in mathematics but with other merits) could escape the routine by entering Ph.D. school ... subject to approval (or being recommended for the other merits) by the Party.

[c] At the end of 1952 a number of medical doctors, the most accomplished medical specialists, were arrested on fabricated charges. All of them, with the exception of two, were Jews. They were accused of acting under the orders of an American Jewish "spy" organization, *Joint*, with intention to kill the leaders of the Communist Party and the Soviet Government under the pretense of healing them. It was only Stalin's death that saved them from capital punishment. Shortly after Stalin's death it was announced that all charges were fabricated. – Editor's note

Social pressure and imprinting of the patterns of behavior of the teachers and parents do play a role in molding the character (hence the fate) of most people. Berezin was a student of I. Gelfand, and used to attend the seminars of L. Landau. Both these geniuses were "strong personalities," not easy to cope with for most. Together with Berezin's mother, they definitely constituted the major part of Berezin's "reference group," as psychologists say. To combine the impact of these personalities with his strife for "honest life" amidst a schizophrenia of the doublespeak and doublethink of the Soviet society (brilliantly depicted by Orwell) was not easy to him.

A somewhat milder, but similar, situation did not, apparently, affect some: Berezin and R. Minlos, another Gelfand's pupil, represented two poles of behavioral pattern.

Minlos was always a jovial personality (at least, outwardly), seemingly incapable of not "thinking positively." Once, after a political blunder he performed (I forgot the details), he was punished: assigned to lead students "to pick potatoes," as this yearly chore was commonly called (officially this test for loyalty and submission to the ruling Party was disguised as help to collective farmers). The majority of the faculty members and some students who knew some details of the story were gossiping: "how terrible!" (professors were usually spared *this* "social work" in lieu of something cleaner, though probably not more palatable), might this be just the beginning of a new purge, etc. I was with Minlos at a conference after which he had to return to Moscow to be sent to this mild version of "hard labor camp"; his reaction to my words of sympathy was: "But this is fine! A month of fresh air and a bit of physical exercise with full salary!!"

Berezin's reactions were sometimes manifestly inadequate. Towards the end of his life, his burdens aggravated. He acquired a family, had to take care of his mother, and desperately tried to keep pace with the explosion of interest in supersymmetry, to hurriedly write up what he could (and what he could not) about the representation theory of Lie superalgebras, leaving several of his most original ideas to be picked up and developed after his death.

Once, less than one year before Berezin's death, we entered the room occupied by the Chair of the Theory of Functions and Functional Analysis, to which he was affiliated (together with Minlos, Palamodov, Kirillov and Gelfand, among more than a dozen of the Chair members). It was in August, the University was empty, and the huge table, large enough to simultaneously accommodate all the faculty members of this large Chair was covered with piles of letters amassed during the vacation period. Some letters were addressed to Kirillov, some to Gelfand, several were addressed to other members of the Chair, but most of the letters were to Berezin. There were too many of them in several piles and some fell on the floor. Several letters landed in a wastepaper basket, obviously by accident or due to a drought. I dived under the table to collect the letters scattered over the floor and fish out the ones from the basket, and was worried by Berezin's sudden pallor. "You see, Dima, how they hate me," he said, and refused to acknowledge the possibility that the letters had been blown off the table.

To me it was clear that, although Berezin was not easy to deal with, and certainly was not liked by several of his colleagues, in this particular case he definitely was purely paranoid. Whether his permanent sense of unbearable stress was justified or not, is debatable. Not many can live with, and take care of, the mother who is from time to time isolated in an institution. Even without such a burden just to live in the Soviet society became gradually impossible to many from all strata of the society.

Berezin was striving to use every spare moment to study, to learn, to do research, to teach, or, when exhausted, engage in some activity in order to be fit to fulfill his self-imposed duties.

Berezin was well aware of his various constraints and the deformities of character they imposed on him. He tried to do his best not to give in. For example, he used to go wandering in the woods. Alone. I found this impossible to understand: Being the son of geologists brought up in a small town in a house with a stove fed on wood and coal, I had some basic accident prevention rules in-built. For example, one may not, under any circumstances, go to the woods or climb mountains alone.

Berezin caused a considerable alarm at a mathematical school held in 1975 in the picturesque foothills of the Northwestern Tian-

Shan mountains when one evening he failed to appear at supper. All night long, some participants and local organizers were wandering along the main road with torches to indicate their location. Berezin appeared early in the morning, having spent the dark and chilly time in the wilderness of a canyon under a bush. Asking for trouble in a similar fashion is a not uncommon way of attempting to liberate oneself from the boring shackles of habitual routine.

For a while, Berezin's clumsiness, which he acknowledged with good humor, kept him away from major disasters. And he tried, desperately, to make fun of his adventures. He boasted to me that once, while he wandered Karelia taiga (finding his path in-between real hard labor camps and villages inhabited by former prisoners, farmers reduced to drunkards and human shambles), he met a desperate looking man who asked if Berezin had a match. This sounded like a regular debut of a mugging, but Berezin was quick enough to produce a condom from his breast pocket, into which a box of matches was carefully rapped to protect them from constant rain and general moisture. "The man respected me," he repeated a couple of times with a chuckle.

I did not want to get contaminated by Berezin's view of life, and tried to prevent this, but I am afraid something contagious stuck in. However, I am thankful to fate even for such a perpetual reminder about Berezin, my teacher, a visionary whose serendipity is coming to be adequately appreciated.

Like Berezin, I try to find an escape from the unbearable by telling stories. Here is one of several I used to tell to my friends about Berezin and his colleagues.

From the history of Berezin's inequalities

"Le soleil entre tes jambes..."
P. Eluard, Les derniers poèmes d'amour [d]

Having amassed a "borderline" score at the entrance exams to the Department of Mechanics and Mathematics of Moscow Univer-

[d]For those who do not read French: "The Sun between your thighs...", P. Eluard, *The latest love poems.*

sity (my name is not Russian, and I graduated from a specialized mathematical school, which in the years following my graduation was counted as next to a crime by the mekh-mat admission committee, see [1]; my school, or rather its organizers, were among the particularly hated ones), I was rejected "in toto." (It was an unjust but a correct move: I knew where I was going and could have prepared better, especially since it was felt that, in a year or two, no matter how one was prepared, it would not help — and it did not, see [1]. So it was a good lesson.)

I was not eager to get conscripted into the Red Army, so to evade a summons, I applied to the department where the faculty of mathematics was, perhaps, the second best after mekh-mat (at least, in Moscow), but where the entrance exams offered no competition at all and, to get admitted, one needed only barely pass: I applied to the "foreign languages" subdivision at the Math Department of the Lenin Moscow State Pedagogical Institute (MGPI). More precisely, to be on the safe side, to the French "stream" (called "Mathematics in French") of this subdivision where the competition was negative[e]. Besides, I liked to learn languages, knew no word in French, and this gave me a chance to study French in earnest and for free.

(As a vitriolic Andrei Zelevinsky commented later, "Mathematics in French... Mmm, and, say, ski in English... I see, I see.")

During the first year of my study in "Ped" as it was commonly called (short for Pedagogical and with a bawdy allusion), I used to attend special seminars at mekh-mat (during the evenings), but a shortage of time was pressing: At the end of the first year, our students' group should have started to actually communicate a bit in real French[f]. We were trained to be teachers in colleges or even universities of the Francophone African countries, friends of the Soviet

[e]At that time, approximately four schoolchildren competed for each available place at mekh-mat of MGU while there were less than needed applicants at the French subdivision of MGPI's mekh-mat. To avoid a looming depopulation, several students particularly dear to the MGPI administration for whatever merits (in their past or future) were added after each exam that weeded out the competing masses.

[f]The goal of studying foreign languages unheard of anywhere in the Soviet Union, except for a few selected places where translators were being trained for military purposes, or the places for the children of the privileged.

Union. (This friendship visibly withered away and our group (1968) was the last.)

The courses in military education during the second year were also more interesting than at mekh-mat: Instead of dull programming, we were taught to drive a lorry (rather irresponsibly as compared with how we were coached in French, but also for free). So I decided not to try to get to mekh-mat for awhile and to stay at MGPI for one more year: Although I had no idea when and where I would be able to drive, I thought that the driver's license — any extra license for that matter — could not hurt.

The general lack of "atmosphere" made the life of those interested in science rather boring.[g] Even military courses (not the drills in marching, which occupied about half of the time, but the driving part) were a welcome distraction.

Another distraction were special courses and seminars at mekh-mat and visits to F. A. Berezin, with whom I got acquainted a couple of years prior to my entrance exams thanks to Berezin's school pal, an uncle of mine. To my chagrin, Berezin did not give me any mathematical assignments at all, instead we chatted on books I had read (both fiction and math), so I was silly enough to think that these were social visits, rather than business. It dawned on me only several years later that Berezin used these discussions of zen-Buddhism or linguistics as a means to understand what I like, in order to offer a problem fitting my inner preferences.

During one such evening, Berezin introduced me to G. L. Litvinov. We left Berezin's flat together and, bidding Berezin farewell, Grigorii Lazarevich suggested we have a walk together, to escort him. Of course I agreed. It was deep autumn, in the evening, sprinkled with occasional drizzle. Grisha[h] was already past twenty, so he posed as a patriarch, and that evening he was even armed with a cane, which,

[g]Even the presence of Bockstein (the discoverer, in the 1930s, of the Bockstein homomorphism and still active enough to jump up on the table from the floor when challenged by his mischievous students; quite impressive for a Professor past sixty and about 150 cm in hight), Raikov and P. S. Novikov could not counterbalance "the social pressure" of the peers to enjoy life rather than waste it on studying any science.

[h]A diminutive of Grigorii, i.e. Gregory.

judging by his gate, he was not in need of at all. Sometimes, to be fair, he used it to measure the depth of the puddles.

Since to speak with me, an analphabetic, about operators of generalized shift on groups — Grisha's forte — was meaningless, Grisha held a monologue, commenting, as is often customary among mathematicians, as well as other people, on various common acquaintances and strangers of interest.

Soon came Berezin's turn.

Once, Grisha recounted, Berezin and he were members of an inspection commission affiliated to the Ministry of Higher and whatever else Education. As Inspectors, they were sent on a business trip to the central Asia. I listened to this story as a fairy tail: Who would grant mathematicians, especially the likes of Berezin or Litvinov, to mooch a free ride (using modern terms) to these exotic places was something beyond my imagination.

At the time of Grisha's story, Berezin studied symbols of operators and had published several papers which at places resembled the treatises of the XIX century, where the formulas of analysts did not fit onto a page, see [2].

Berezin and Litvinov shared a hotel room, and Litvinov was intrigued beyond reason by Berezin's behavior: during the day they performed their duties as Inspectors, and on the rare evening that was free of "friendly dinners" lavished on them by those they inspected, Berezin assiduously studied some booklet carefully wrapped so that its title was hidden from Grisha's inquisitive eyes. After a while, having tucked the booklet in his jacket pocket, Berezin used to sit at the table and, sometimes peeping inside the booklet, wrote longish formulas.

Although rewriting, preferably without due references, other people's papers (especially from a foreign language) was experiencing a fresh upsurge at that time, Grisha could not even think that Berezin was just reproducing (even with improvements) somebody's foreign preprint. (The format of the booklet was, evidently, foreign.) Berezin manifestly did not want to show the booklet and skillfully avoided all discussions of it.

By this time of Grisha's narrative I was a bit tired to listen about the mysterious (to a freshman) symbols of operators or mathematicians' traditions of relations to intellectual property (which I soon had to experience first hand) and was in a hurry to get home where my first book of French verse was waiting for me. Although it was in the original (a collection of Eluard's poetry) it was understandable. Certain bits I had already deciphered were rather inflammatory; I was eager to continue.

And as I was in the middle of navigating through a most slippery chain of puddles, Grisha confessed that once, when Berezin went out for a moment, Grisha could not stand the uncertainty any longer. Ignoring his Mother's (and Grandmother's) instructions on behavior and his un-sportive disposition alike, he jumped to the table, took the booklet out of Berezin's jacket and looked inside, all a-trembling. Grisha told me that, actually, he did not know French, except a bit, sufficient to read mathematics. But, although the booklet was in French and not mathematical, he could master the title: "P. Eluard. *Les dernier poèms d'amour.*"

This story added to my, already considerable, eagerness to become Berezin's student. I never regretted.

References

1. M. Shifman (ed.) *You Failed Your Math Test, Comrade Einstein,* (World Scientific Publishing, Singapore, 2005).
2. F. A. Berezin, *Convex Functions of Operators*, (in Russian) *Mat. Sb. (N.S.)* 88 (130) (1972), 268–276.

PART III

From Felix Berezin's Archives

Letter to Academician R.V. Khokhlov,
the Rector of the Moscow State University

Letter to the Governing Board of the Moscow
Mathematical Society

Letter to Academician R.V. Khokhlov, the Rector of the Moscow State University [a]

Esteemed Rem Viktorovich,

I deem it my duty to impart several of my considerations concerning the mathematical department of mekh-mat.[b] In my opinion, its present state is unsatisfactory and its future is alarming.

There is a group of professors on the faculty of mekh-mat, prominent mathematicians, whose activities represent the scientific face of the faculty, at least its mathematical part. It seems natural that these people should take active steps in the decisions involving vital questions regarding mekh-mat: the acceptance of new students and Ph.D. students, affairs of promotion, and the direction of scientific policies. The present administration of mekh-mat, however, practically completely barred the most esteemed mathematicians from participating in any of these questions. The decisions are made by a small circle of people, often unqualified; the members of the faculty who do not enter this narrow group have no means of influencing them. As a result, a huge damage is inflicted on mekh-mat.

In my opinion, if the present situation does not change in a all-out way, there is a danger of degrading mekh-mat as the all-Union[c] center for the preparation of highly qualified mathematicians and as a scientific center.

I would like to give several facts from various areas of the mekh-mat's life that confirm this opinion.

[a] Translated from Russian by Roman K. Kovalev, The College of New Jersey, Department of History, Ewing, NJ 08628, USA; e-mail: kovalev@tcnj.edu. The original is published in *Zapiski Nauchnykh Seminarov POMI*, v. 331, 2006. French translation in *La Gazette des Mathématiciens*, le bulletin interne de la Société Mathématique de France, **110** , 47, Octobre 2006.

[b] Faculty of Mechanics and Mathematics of the Moscow State University –Translator's note

[c] In the USSR the adjective "all-Union" was used in the meaning of "national." —D. Leites' note

I. Acceptance of new students

This is the principle question, vital to the life of the mekh-mat. At this moment, the public interests shift away from the natural sciences to the humanities, which, in particular, reduces the competition at the entrance examinations to the mekh-mat. On the other hand, the well-developed system of tutors masks the difference between applicants who have been coached for examinations to solve typical problems and applicants who have natural mathematical talent but were not trained. In such a situation, responsible attitude to the entrance examinations attains a particular significance.

From the moment of inauguration of the current administration, the entrance examinations are guided by a vice chairman, M.K. Potapov. During his tenure, none of the well-known mathematicians, even the ones well acquainted with the school curriculum, took part in the Admission Committee. Among such experts are, for example, Professors V.M. Alekseev, V.I. Arnold, A.A. Kirillov, V.P. Palamodov, Docents[d] A.M. Stepin, and M.A. Shubin. This list is far from being complete. As a result of the present unqualified activity of the Admission Committee, the average level of students had steeply declined, and the number of mathematically talented students amongst them also plummeted.

The lowering of the average level has been noted by various instructors and professors, who have been on the faculty long enough to be able to compare the present state with the past. This decrease is not reflected in achievement scores, since at the examinations a "renormalization" to the average level of the students always takes place. This renormalization sometimes takes very precise and characteristic forms.

Thus, for instance, at the Chair of Differential Equations, in an attempt to raise the level of academic achievements, examinations have been made simpler: if previously each examination consisted of a theoretical question and a problem, now there is no problem.

The diminished number of talented mathematicians among the students is reflected in the quality of Ph.D. students. It defies

[d]Equivalent to Associate Professor in the USA. –D. Leites' note

a formal quality control, but many members of the faculty feel it. Apparently, the diminished number of talented mathematicians among the students is not a secret to the Administration. This can be inferred from a graphic episode that had occurred in the past year at the Students Mathematical Olympiad, which was organized by our department. According to the rules of the Olympiad, all Institutions of Higher Education[e] are divided into three groups so that mekh-mat belongs to the first group while MIIT (Moscow Institute of Transport Engineers), among other (poly)technical Institutes, falls into the second group. Last year, the MIIT team took first place in its group, having by far outdone all its rivals. Based on this, MIIT requested to be transferred to the first group this year; the request was, however, denied. Be that as it may, the students — members of the MIIT team — asked auditorium proctors for the texts of the problems assigned for the members of the first group and proceeded to solve them. When this was found out, the MIIT team was disqualified.

The only rational explanation of this, that comes to my mind, is as follows. The Administration is well aware of the fact that MIIT accepted a number of very able mathematicians who were denied admission to mekh-mat. Their participation in the team rival to ours could well have deprived our team of its victory or, in any event, make the victory less convincing.

One can, to some extent, judge the nature of the work done by the Admission Committee by the following fact. I know eleven names of the winners of Moscow and National School Mathematical Olympiads who were not accepted to mekh-mat in 1975. This list is definitely incomplete.

During the course of many years mekh-mat nourished a tradition of actively recruiting the students. This recruiting has been carried out on various levels: through school mathematical circles; by organizing Moscow Mathematical Olympiad and satellite Olympiads in smaller towns; and, finally, by transfer to the second or third year of mekh-mat the students from provincial (regional) universities, after

[e]In the US and other western countries "institutions of higher education" are usually referred to as universities. –D. Leites' note

they had completed the introductory schooling. Our faculty members were intentionally sent to these provincial Universities in order to pick up candidates for the transfer.

At present, there is a generation of important mathematicians whose fates evolved in this way. Among them I would like to point out J. Khadzhiev, my student, who has recently defended his Doctorate dissertation, and is now the Chairman of the Mathematics Department at the University of Tashkent.

At present, this recruiting is abandoned. Incidentally, it should be noted that recently, in response to the Rectorate's decree, the Evening Mathematics School of MGU[f] was revitalized. However, it had been in existence in its new state only for a short time, so it is too early to judge its role in attracting new students.

The trips of faculty members to regional universities with the purpose of recruiting junior students as candidates for transfer to mekh-mat completely terminated. This circumstance sharply reduces the role of mekh-mat as a National center for preparation of highly-qualified mathematicians.

During the past year, no one of the faculty members was sent to carry out regional Mathematical Olympiads.

II. Admission to the Graduate School

During the tenure of the present Administration, for a wide range of strange reasons, a number of students recommended by their scientific advisors, were not permitted to take the entrance examinations to Graduate School. Some of these people — Khovanskii, Blekher, Zarkhin, Koitman — soon defended their theses outside the MGU Graduate School[g] and are at the moment recognized as mathematicians of considerable merit.

Such situations were especially common recently at the Department of Algebra. Kifer and Skorniakov (both recommended

[f]Moscow State University. –Translator's note

[g]In the Soviet Union it was theoretically possible to approach a scientific council "from the cold" and present a thesis without being officially guided by a scientific advisor. In the above cases, however, one had to have an (influential) advisor who would "help" to organize the defense. —D. Leites' note

by Yu.I. Manin) and Leites (recommended by A.L. Onishchik) were not allowed to take the entrance examinations to Graduate School. Concerning Leites, I know that at the moment he graduated from a pedagogical university, he had already written several papers, one of which was published while the rest were submitted for publication. He was even a rather active member of *Komsomol.*[h] The reason given for his rejection was based on the fact that he received a "3"[i] in his sophomore year. In light of the present circumstances, the reason offered for his rejection seems purely formal.

III. Administration's treatment of the faculty. Admitting new students

I believe that the administration of mekh-mat irresponsibly approaches the hiring of new faculty members. For reasons that have nothing to do with his scholarly abilities, the brilliant mathematician Gabrielov was denied employment; several years ago, he graduated from the Graduate School under the supervision of Professor V.P. Palamodov (the Chair of Theory of Functions and Functions Analysis). At the same time, a number of people without necessary qualifications were hired. There has recently developed a trend of giving permanent faculty positions by purely administrative fiat, i.e., without any discussion of the candidates with the members of the Chair to which these candidates are hired. (For example, in 1976, A.A. Shkalikov was hired at the Chair of Theory of Functions and Functional Analysis as soon as he completed the Graduate School, and he was not introduced to the faculty members of the Chair.) Such a practice conceals a very grave danger of polluting the faculty with unacceptable cadres.

Administration's policy regarding "moonlighters" is bewildering. It seems obvious that one should invite only the greatest mathematicians of the Soviet Union to hold a joint appointment, or specialists in the branches of mathematics underrepresented at mekh-mat. This, however, is not always the case. For example, at the Chair

[h] Young Communist League. –Translator's note
[i] Equivalent to a "C." –Translator's note

of the Theory of Functions, among the holders of joint appointments we see I.M. Gelfand, A.V. Vitushkin, A.A. Gonchar, S.B. Stechkin, and A.K. Gushchin. I believe that the offer to Stechkin and Gushchin is fully unwarranted. Stechkin is a famous mathematician and a good lecturer. Nevertheless, he is an expert in the theory of functions which is very well represented at our Chair by the main members and scientifically does not stand above them in any significant way. As for Gushchin, he is a rather secondary scholar in differential equations. In other words, not only is his scientific level inadequate, in my opinion, for offering him a joint appointment, but even his mathematical expertise has no relevance to our Chair.

The mekh-mat Administration humiliates faculty members by treating many without due respect. One of the examples of such an attitude is the failure to invite several prominent mathematicians who work at the department, such as Professors A.A. Kirillov and V.P. Palamodov, to take part in the scientific committees of the department.

Another example is a common practice, lately, of delaying decisions of promotion and tenure[j] for a year or more. Among the faculty who found themselves in such situations there is, for instance, a talented mathematician, Docent A.M. Vinogradov. His recertification has been delayed over the course of the last two years. The most disgraceful of such stories is, however, the one concerning Professor G.I. Shilov, who was not recertified for the next term and died shortly after. Shilov was not only a significant mathematician, whose fate was inseparable from that of mekh-mat, but also a brilliant pedagogue, the author of very popular textbooks, and one of the creators of the course on functional analysis at our department (the so-called "Analysis III"). Shilov's services to our department are very

[j]The nominally permanent positions in Soviet universities and research institutes of the Academy of Sciences were subject to "recertification" every four years or so. One could be exempt from such recertification for outstanding social activities, if approved by the Party, even if one had no publications for years or decades. A postponement of the decision was usually considered as a hint before firing unwanted scholars, to give them time either to find a new job or to demonstrate that they learned "how to behave." —D. Leites' note

great and his absence will be felt for a long time to come.

Unwilling to tolerate further neglect from the Administration, a number of the faculty members have resigned from the department; for example, a major mathematician who led active pedagogical work at the department, Doctor of Sciences, A.L. Onishchik. If the situation at the department does not improve, several other faculty members, who presently form a scientific core of the department, may follow the suit.[k]

IV. Scientific policy

According to the stipulations of the mathematics branch of our Department, the Administration has no right to determine the nature of the scientific subjects chosen for study by the faculty. For this reason, scientific policy can only consist in inviting scientists from the outside to reinforce certain scientific direction and in establishing contacts with various scientific centers.

In my opinion, the scientific policy of the current administration of our department is unsatisfactory. First of all, I shall refer to the example noted earlier. Despite the fact that the Chair of the Theory of Functions is very well represented at mekh-mat, the expert in this field, S.B. Stechkin, was invited to take a joint appointment position. At the same time, although the MGU mathematical physics is not in such a fine shape at all, theoretical physicists recruited by I.G. Petrovskii[l] to strengthen this direction, are being fired.

Moreover, it was prohibited to Professor A.S. Schwarz, one of the leading specialists in the Soviet Union in the area of mathematical aspects of quantum field theory, to deliver his special course on quantum field theory. This special course was taught on a volunteer basis (!); it was interrupted in the middle of the academic year.

In the same way the Administration terminated employment of the well-known mathematicians, experts in mathematical economics,

[k]In fact, that's exactly what happened, even before the mass emigration started. –D. Leites' note

[l]A talented mathematician, a member of many Academies, and a shrewd politician, the former Rector of the Moscow State University, NOT a member of the Communist Party. –D. Leites' note

permanent members of TsEMI,[m] B.S. Mityagin and A.B. Katok,[n] despite the fact that mathematical economics is unsatisfactorily represented in our Department. In this way were destroyed the seedlings of scientific contacts between our faculty and TsEMI.

In general, among the traditions of our Department, there has always been a very free atmosphere of active participation of mathematicians, who are not members of the faculty, possibility to lead seminars or specialized courses (either without pay or on hourly payments). This possibility always raised the tonus of work with students and served as a versatile method for strengthening the scientific directions that required it. The current Administration is the first to regulate such a practice, without, incidentally, coordinating this regulation in any way with the scientific value of the program or with the popularity among the students attending the special course or special seminar.

Finally, I would like to note that over the past years a number of major international scientific centers, such as the Institute of Advanced Studies (Princeton, USA), Oxford University (United Kingdom), and CERN (Switzerland), expressed considerable interest in developing contacts with our department, having invited several members of our faculty.[o] The administration of the department, however, for various far-fetched reasons, prevented these contacts. I believe that this had a detrimental impact on the development of mathematics in our country.

I would like to take the opportunity here to express my thoughts on the necessary ways of how to improve the work of the department.

1. *Major scientists should stand at the helm of the Department*, those whose professional standing is significant. Above all, this pertains to the chairperson. At the moment, it seems to me, there is no one who could equally competently head the mechanics and mathematics branches of mekh-mat. Perhaps, because of this, it

[m] Central Economics-Mathematics Institute. –D. Leites' note

[n] Both are now prominent professors in the USA. –D. Leites' note

[o] F.A. Berezin does not indicate that these three invitations were made specifically to him, but it is most likely that he was not alone to receive such invitations. – E.G. Karpel's note.

is advisable to introduce the position of vice-chairperson in charge of the Mechanics branch if the Chairperson of mekh-mat is a mathematician, and vice-chairperson for the Mathematics branch, if the Chairperson of mekh-mat happens to be an expert in mechanics.

2. *I believe that one of the reasons why mekh-mat is presently functioning in an unsatisfactory way* is because there is an impenetrable layer of secrecy which envelops all the actions of the current Administration. This issue, in my opinion, is central. The activities of administration should be open.

To achieve this goal, the following would be useful:

a. The annual reports of the Administration have to cease being a pure formality. For this to work, the administration must dispatch in good time, no later than two weeks ahead, the text of the report or at least its summary.

b. Minutes of every meeting held by the Chairperson should be posted, which should include a roster of the agenda, the issues considered, the outcome of discussions, and a list of those present.

These minutes should be filed; members of the department should have access to all of them upon request; and, the Administration must not make any decisions not mentioned in the bulletins of these meetings.

c. When a new faculty member is considered for hire, it should be necessary to have a preliminary meeting of the faculty members of the Chair to which the new member is being hired. At this meeting, the head of the Chair or a representative of the Administration has to present a report on the scientific merits of the candidate. It is advisable to invite to such a meeting members of other Chairs with common scientific interests with the candidate.

d. Scientific councils should discuss all the candidates recommended for Graduate School by the faculty members. If a candidate invites objections, these objections have to be expressed openly.

3. *It is necessary to improve the composition of scientific councils.* As I have already wrote above, at the moment, a number of authoritative scientists who comprise the main body of the faculty are not taking part of these councils.

At the same time, the scientific councils are manned by strange members who systematically file "null and void" reports when voting at Ph.D. or Doctoral defenses or vote "for" or "against" regardless of the quality of the dissertation. As a result, twice this spring, shameful situations occurred when, having only the positive oral and written reports, the scientific councils flunked the dissertations. (One member of the scientific council even made an attempt to justify such behavior on the part of the council by evoking the secret nature of the vote.) In both cases, in an attempt to avoid a more significant scandal, the scientific councils recognized the protocol of tallying committee "invalid" so as to be able to vote for the second time.

4. *I believe that it is necessary to restore* the long-held tradition that every actively working mathematician, even not a member of the faculty, may lead a special seminar or read a special course either without pay or for hourly rates. No formalities, except for permission of the interested Chair, should be required for the teaching of the class.

5. *I believe that in order to have a full-blooded life* at the Department of Mathematics, it is necessary to advance not only the traditional directions of internal mathematical significance, but also other scientific disciplines which have mathematics at their base.

To me, the most significant of these are:

a) Mathematical physics in the broad sense of the word, which includes mathematical foundations of quantum mechanics, statistical physics, and quantum field theory;

b) Mathematical biology;

c) Mathematical economics.

Currently, the situation with mathematical physics looks rather satisfactory. Mathematical biology and mathematical economics are very poorly represented at the Department. In Moscow, there are several very strong mathematical communities working in mathematical biology and mathematical economics. I think that it is advisable to encourage their cooperation with our faculty.

In the future, it seems to me, it would be sensible to establish Chairs of Mathematical Physics, Mathematical Biology, and Mathematical Economics.

The evaluations of the activities of the Administration and certain mathematicians given in this letter are my own, and therefore, perhaps, subjective. However, I know that these observations are shared by a number of other faculty members.

As for the various facts mentioned here, they can also be testified by several faculty members.

F.A. Berezin
Department of Mechanics and Mathematics
Chair of Theory of Functions and Functional Analysis

To the Governing Board of the Moscow Mathematical Society [P]

February 17, 1970

Over the last several years, conflicts over the defense of dissertations have increased. The reason for this, partly, stems from the discrepancies in the level of expectations of the dissertations, and, partly, due to the unfortunate reasons unrelated to mathematics.

I think that the Moscow Mathematical Society should not stay outside these conflicts, at least, in cases where the dissertations are those of the members of the Society.

In my opinion, whenever a member of the Society is preparing to defend a Doctoral thesis, the Governing Board of the Society should have a firm opinion about the quality of the work. If the Governing Board believes that the thesis proposed satisfies the necessary requirements, it should support it; and, if it does not, then it should recommend the author of the corresponding thesis to withhold the thesis from submission to defense. The support of a thesis may, by a decision of the Governing Board, take a number of forms, but it is imperative that a part of it is formation of a definite public opinion. (In addition to this, there are other ways of showing support, such as presentations at the defense or, if necessary, an official inquiry to the Higher Attestation Committee.)

If a conflict arises before the defense, it seems sensible to me to appeal to the experts in the same branches of mathematics to which the given thesis belongs, chosen by mutual agreement, to determine the nature of the conflict. Thereafter, a meeting should be held with all the interested parties for a detailed discussion. In some cases, a report of such a meeting should, perhaps, be published in the Russian Mathematical Surveys in the section "Mathematical Life in the USSR."

The goal of such a discussion is for the public to better understand the quality of the thesis proposed and learn about the nature of the

---

[P] Translated from Russian by Roman K. Kovalev, The College of New Jersey, Department of History, Ewing, NJ 08628, USA; e-mail: kovalev@tcnj.edu.

objections made against it. I think that the initiative I suggest to the Moscow Mathematical Society will assist in establishing generally recognized criteria of the level of works that can be defended as Doctorate theses (DSc) and would be of great moral significance. This initiative would also assist in elevating the prestige of the Moscow Mathematical Society.

F.A. Berezin

I said, and thus saved my soul.[q]

[q]Latin expression *Dixi et animam levavi* – "I have spoken and relieved my soul" – here is intentionally paraphrased and emphasized. –E.G. Karpel's note

From Wikipedia, the free encyclopedia:

Felix Alexandrovich Berezin (Russian: Феликс Александрович Березин) (25 April 1931 – July 14, 1980) was a Soviet mathematician and mathematical physicist known for his contributions to the theory of supersymmetry and supermanifolds as well as to the path integral formulation of quantum field theory. Berezin studied at the Moscow State University, but was not allowed to do his graduate studies there on account of his Jewish origin (his mother was Jewish). After graduation for the next three years Berezin taught at Moscow high schools. He continued to study mathematical physics under the direction of Israil Gelfand. After Khrushchev's liberalization he joined the Department of Mechanics and Mathematics (mekh-mat) at the Moscow State University at the age of 25.